ORDINARY DIFFERENTIAL EQUATIONS AND STABILITY THEORY

AN INTRODUCTION

ORDINARY DIFFERENTIAL EQUATIONS AND STABILITY THEORY

AN INTRODUCTION

by

David A. Sánchez
University of New Mexico

Dover Publications, Inc.
New York

Published in Canada by General Publishing Company, Ltd., 30 Lesmill Road, Don Mills, Toronto, Ontario.
Published in the United Kingdom by Constable and Company, Ltd.

This Dover edition, first published in 1979, is an unabridged and unaltered republication of the work originally published in 1968 by W. H. Freeman and Company. The Dover edition is published by special arrangement with W. H. Freeman and Company, 660 Market St., San Francisco, Calif. 94104.

International Standard Book Number: 0-486-63828-6
Library of Congress Catalog Card Number: 79-52007

Manufactured in the United States of America
Dover Publications, Inc.
180 Varick Street
New York, N.Y. 10014

Preface

In this book I have attempted to give a brief, modern introduction to the subject of ordinary differential equations, with an emphasis on stability theory. This emphasis has been directed chiefly toward the undergraduate or beginning graduate student, who is often deprived of any exposure to the newer concepts.

Neglect of these concepts is usually justified by the argument that the student is insufficiently equipped mathematically to handle them. As a result, the usual course in ordinary differential equations consists in learning a battery of special techniques to solve special equations; possibly an existence theorem is proved.

This is an injustice to the student, for in actuality a modicum of knowledge beyond the calculus will carry him a relatively long way in the study of stability theory. If his appetite is whetted, it should serve as an incentive to equip himself to proceed further. In any case the student will be aware that the finding of solutions of exotic ordinary differential equations is not the principal aim of the subject.

Chapters 1, 2, 3, and 6 consider the usual problems of existence and uniqueness of solutions, the maximum interval of existence, fundamental systems of solutions of linear equations, and nonhomogeneous linear equations. I chose to begin with a discussion of the first-order linear system for two reasons. First of all, the results for the nth-order linear equation follow with almost no effort. Second, and most important, the notion of a fundamental matrix is developed.

This is used in the representation of solutions of homogeneous and inhomogeneous systems, and plays a key role in the discussion of stability of nonautonomous systems.

Chapters 4 and 5 are introductory discussions of stability theory for autonomous and nonautonomous systems. I have made no attempt to be encyclopedic, but merely give some basic results. These are a matter of personal choice and of the intended audience, but hopefully they will serve the ambitious reader as a suitable starting point in this vast field.

The problems given are, for the most part, designed to fill out the text material. Appendix A deals with series solutions near regular singular points. The proof of the main theorem requires some knowledge of complex variables; it may be omitted and the reader may proceed directly to the examples given. Appendix B gives some results dealing with periodic solutions. The books listed as references were extremely useful to the author, and the reader will be well rewarded in investigating their contents.

I wish to thank Professors K. S. Williams and J. D. Schuur, Doctor J. L. Brenner, and Miss Patrice Whittlesey for their comments and suggestions. My appreciation is extended to the Mathematics Departments of the University of Manchester and of the University of California at Los Angeles for technical assistance, and to Mrs. Audrey Biggar and Mrs. Ruth Goldstein for typing. Finally I wish to thank the undergraduates who endured my preliminary classroom efforts to give ordinary differential equations a modern meaning.

D. A. SÁNCHEZ

Los Angeles, California
May 1967

Contents

ORDINARY DIFFERENTIAL EQUATIONS AND STABILITY THEORY

AN INTRODUCTION

CHAPTER

1

Introduction

1.1 Preliminary Notation

For convenience, we will employ vector notation throughout the text, and n-dimensional vector-valued functions of n- or $(n + 1)$-dimensional vectors will be most frequently used. Thus if the positive integer n is unspecified, then any results stated will be applicable to one- or many-dimensional problems.

In general, suppose we are given the function f mapping a subset of R^m, Euclidean m-dimensional space, into R^n, Euclidean n-dimensional space. If $x = (x_1, \ldots, x_m)$ is in the domain of f, and we denote its image under f by $y = (y_1, \ldots, y_n)$, then we may write

$$y = f(x) = (f_1(x), \ldots, f_n(x)),$$

where we define

$$y_i = f_i(x) = f_i(x_1, \ldots, x_m), \qquad i = 1, \ldots, n.$$

We will say that f is continuous in x if each f_i is continuous in x. Furthermore, we define the vector of partial derivatives as

$$\frac{\partial f}{\partial x_j} = \left(\frac{\partial f_1}{\partial x_j}, \ldots, \frac{\partial f_n}{\partial x_j}\right),$$

where $1 \leq j \leq m$.

Given the n-dimensional vector $x(t) = (x_1(t), \ldots, x_n(t))$, where t is a real variable and each $x_i(t)$ is real-valued, we say $x(t)$ is continuous at $t = t_0$ if each $x_i(t)$ is continuous at $t = t_0$, and it is differentiable if each $x_i(t)$ is differentiable. We then may express the derivative vector as

$$\dot{x}(t) = \frac{dx}{dt} = (\dot{x}_1(t), \ldots, \dot{x}_n(t)),$$

and successive derivatives will be denoted by $\ddot{x}(t), x^{(3)}(t), \ldots, x^{(k)}(t)$.

If $x(t)$ is given as above, we denote the norm of $x(t)$ by

$$\|x(t)\| = \sum_{i=1}^{n} |x_i(t)|,$$

and for each t this is a mapping of $x(t)$ into the nonnegative real numbers. It has the properties (i) $\|x(t)\| = 0$ if and only if $x(t) = 0$, that is, each $x_i(t)$ is zero; (ii) $\|kx(t)\| = |k|\|x(t)\|$ for any real or complex scalar k; and (iii) $\|x(t) + y(t)\| \leq \|x(t)\| + \|y(t)\|$. The above norm has certain computational advantages over the usual Euclidean norm,

$$\|x(t)\| = \left\{\sum_{i=1}^{n} |x_i(t)|^2\right\}^{1/2},$$

which also satisfies the characteristic properties of the norm listed above. For geometrical convenience (for example, when using polar coordinates) we will occasionally use the latter norm; any results given will not, however, depend on the norm chosen.

Frequently in the text we will be considering a given function f mapping a subset of R^{n+1} into R^n. If we denote a point in R^{n+1} by (t, x), where t is real and $x = (x_1, \ldots, x_n)$, then its image, wherever defined, may be denoted by

$$y = (y_1, \ldots, y_n) = f(t, x) = (f_1(t, x), \ldots, f_n(t, x)).$$

In particular, if $x = x(t) = (x_1(t), \ldots, x_n(t))$, then $y = y(t) = f(t, x(t))$ is an element of R^n dependent on the real variable t.

If $f(t, x(t))$ is continuous for (say) $t_1 \leq t \leq t_2$, then we can define the integral

$$\int_{t_1}^{t_2} f(t, x(t))\, dt = \left(\int_{t_1}^{t_2} f_1(t, x(t))\, dt, \ldots, \int_{t_1}^{t_2} f_n(t, x(t))\, dt\right),$$

and the usual rules of integration will hold.

Example: Since

$$\left|\int_{t_1}^{t_2} f_i(t, x(t))\, dt\right| \le \int_{t_1}^{t_2} |f_i(t, x(t))|\, dt \text{ for } i = 1, \ldots, n,$$

then

$$\sum_{i=1}^{n} \left|\int_{t_1}^{t_2} f_i(t, x(t))\, dt\right| \le \sum_{i=1}^{n} \int_{t_1}^{t_2} |f_i(t, x(t))|\, dt,$$

which is equivalent to the statement

$$\left\|\int_{t_1}^{t_2} f(t, x(t))\, dt\right\| \le \int_{t_1}^{t_2} \|f(t, x(t))\|\, dt.$$

We will also consider complex-valued functions $z(t) = u(t) + iv(t)$, where u and v are real-valued functions of the real variable t. Then we say that $z(t)$ is continuous at $t = t_0$ if $u(t)$ and $v(t)$ are continuous at $t = t_0$, and $z(t)$ is differentiable if $u(t)$ and $v(t)$ are differentiable. The derivative of $z(t)$ is given by $\dot{z}(t) = \dot{u}(t) + i\dot{v}(t)$, and the usual roles of differentiation apply. In addition we define the usual complex modulus:

$$|z(t)| = [u^2(t) + v^2(t)]^{1/2}.$$

By a scalar function we will always mean a real- or complex-valued function of the real variable t.

Example: Let $x = (x_1, x_2)$ belong to R^2, and then define the function f mapping R^3 into R^2 by

$$y = f(t, x) = (tx_1x_2, 3t^2x_1 + x_2).$$

Hence

$$y_1 = f_1(t, x) = tx_1x_2, \qquad y_2 = f_2(t, x) = 3t^2x_1 + x_2.$$

If $x_1(t) = t$ and $x_2(t) = \cos t$, then $x(t) = (t, \cos t)$ and $y(t) = f(t, x(t)) = (t^2 \cos t, 3t^3 + \cos t)$. Then, for instance,

$$\dot{y}(t) = (2t \cos t - t^2 \sin t, 9t^2 - \sin t)$$

and

$$\int_0^{\pi/2} y(t)\, dt = \left(\frac{\pi^2}{4} - 1, \frac{3\pi^4}{64} + 1\right).$$

If, instead, $x_1(t) = t$ and $x_2(t) = t + i \sin t$, then

$$x(t) = (t, t + i \sin t)$$

and

$$\begin{aligned} y(t) &= f(t, x(t)) \\ &= [t^3 + it^2 \sin t, (3t^3 + t) + i \sin t]. \end{aligned}$$

In this case, for instance,

$$\dot{y}(t) = [3t^2 + i(2t \sin t + t^2 \cos t), (9t^2 + 1) + i \cos t].$$

Finally, we will not distinguish between the scalar 0 and the zero vector having zero as each of its components. The meaning will always be clear from the context.

1.2 The Ordinary Differential Equation

The subject of study is an ordinary differential equation—an equation containing the derivatives of an unknown function $x(t)$, with t a real variable, and possibly containing the unknown function itself, the independent variable t, and given functions. In addition, initial conditions, which the unknown function is required to satisfy, may be given. With such an equation, the object is two-fold: (i) to find the unknown function or class of functions satisfying the equation, and (ii) whether (i) is possible or not, to gain some information about the behavior of any function satisfying the equation.

DEFINITION: The order of a differential equation is the order of the highest derivative of the unknown function appearing in it.

Therefore the general form of an ordinary differential equation of kth order is

$$F(t, x, \dot{x}, \ldots, x^{(k)}) = 0, \tag{1}$$

where $x = x(t) = (x_1(t), \ldots, x_n(t))$ is an unknown function, and F is a function defined on some subset of $R^{n(k+1)+1}$.

DEFINITION: A function $x = \varphi(t) = (\varphi_1(t), \ldots, \varphi_n(t))$, $r_1 < t < r_2$, which when substituted in (1) reduces it to an identity, is called a solution of (1), and (r_1, r_2) is its interval of definition.

Since we assume that $k \geq 1$, it follows that $\varphi(t)$ is differentiable and hence continuous on $r_1 < t < r_2$. If the domain of the function F is some set B in $R^{n(k+1)+1}$, then the point $(t, \varphi(t), \dot{\varphi}(t), \ldots, \varphi^{(k)}(t))$ belongs to B for $r_1 < t < r_2$.

Very little can be said about the equation in the form given in (1), so let us assume that we can solve (locally) for $x^{(k)}$. We obtain

$$x^{(k)} = G(t, x, \dot{x}, \ldots, x^{(k-1)}), \tag{2}$$

the kth-order equation in *normal* form. In this case, since $x^{(k)}$ is an n-dimensional vector, the function G is a mapping from some subset of R^{kn+1} into R^n.

Examples

(*a*) $\dot{x} = a(t)x + b(t)$, where $a(t)$ and $b(t)$ are given scalar functions and $x = x(t)$ is an unknown scalar function, is a first-order equation, and $G(t, x) = a(t)x + b(t)$ is a mapping from R^2 into R^1.

(*b*) $x^{(4)} = [a(t)x^{(3)} + b(t)x^2]^{1/2}$, where $a(t)$ and $b(t)$ are given scalar functions and $x = x(t)$ is an unknown scalar function, is a fourth-order equation, and $G(t, x, \dot{x}, \ddot{x}, x^{(3)}) = [a(t)x^{(3)} + b(t)x^2]^{1/2}$ is a mapping from R^5 into R^1.

(*c*) The system

$$\begin{aligned}
\dot{x}_1 &= tx_1 + 2x_2 + x_3, \\
\dot{x}_2 &= \qquad\quad t^2x_2 + (x_3)^2, \\
\dot{x}_3 &= 5x_1 + 2x_2 + t^3x_3 + e^t,
\end{aligned}$$

where $x = x(t) = (x_1(t), x_2(t), x_3(t))$ is an unknown vector function in R^3, is a first-order equation, and

$$\begin{aligned}
G(t, x) &= (G_1(t, x), G_2(t, x), G_3(t, x)) \\
&= (tx_1 + 2x_2 + x_3, t^2x_2 + (x_3)^2, 5x_1 + 2x_2 + t^3x_3 + e^t)
\end{aligned}$$

is a mapping from R^4 into R^3.

It should be noted that an equation of the form (2) with $n > 1$ is sometimes called a kth-order system.

Given an equation in the form (2) with $k > 1$, the following substitution reduces it to an equation with $k = 1$. Let $y_1 = x$ and $y_2 = \dot{x}, \ldots, y_k = x^{(k-1)}$; then $\dot{y}_1 = \dot{x} = y_2, \dot{y}_2 = \ddot{x} = y_3, \ldots, y_{k-1} = x^{(k-1)} = y_k$, and $\dot{y}_k = x^{(k)} = G(t, x, \dot{x}, \ldots, x^{(k-1)}) = G(t, y_1, y_2, \ldots, y_k) = G(t, y)$. This is the first-order system $\dot{y} = f(t, y)$, where $y = (y_1, \ldots, y_k)$ and $f(t, y) = (f_1(t, y), \ldots, f_k(t, y)) = (y_2, \ldots, y_k, G(t, y)$. Note that if x is an n-dimensional vector, then y is a $(k \times n)$-dimensional vector.

Examples

(*a*) Given any second-order equation $\ddot{x} = g(t, x, \dot{x})$, where $x = x(t)$ is an unknown scalar function, and letting $y_1 = x$, $y_2 = \dot{x}$, we have the system $\dot{y}_1 = y_2$, $\dot{y}_2 = g(t, y_1, y_2)$. If we let $y_2 = y$, this can be written, $\dot{x} = y$, $\dot{y} = g(t, x, y)$, a first-order equation, where $(x, y) = (x(t), y(t))$ is an unknown two-dimensional vector. In the first equation a solution would be a scalar function; in the second equation it is a pair consisting of a scalar function and its first derivative.

(*b*) The second-order system $\ddot{x}_1 = ax_1 + b\dot{x}_2$, $\ddot{x}_2 = c\dot{x}_1 + dx_2$, where $x(t) = (x_1(t), x_2(t))$ is an unknown two-dimensional vector, is transformed as follows. Let $(y_1, y_2) = x = (x_1, x_2)$ and $(y_3, y_4) = \dot{x} = (\dot{x}_1, \dot{x}_2) = (\dot{y}_1, \dot{y}_2)$. Then we have the first-order system

$$\dot{y}_1 = y_3, \qquad \dot{y}_2 = y_4, \qquad \dot{y}_3 = ay_1 + by_4, \qquad \dot{y}_4 = cy_3 + dy_2,$$

where $y = y(t) = (y_1(t), y_2(t), y_3(t), y_4(t))$ is an unknown four-dimensional vector function.

It follows that we need only consider the first-order equation

$$\dot{x} = f(t, x), \tag{3}$$

where $x = x(t)$ is an unknown n-dimensional vector function, and $f(t, x)$ is a mapping from a subset of R^{n+1} into R^n.

Given the first-order equation (3), let us assumed that $f(t, x)$ is defined in a domain B, an open connected set in R^{n+1}. If $f(t, x)$ is continuous in B, and $x = \varphi(t) = (\varphi_1(t), \ldots, \varphi_n(t))$, $r_1 < t < r_2$, is a solution of (3), then it may be thought of as a curve lying entirely in B. Furthermore, it will have a continuously turning tangent at each point given by

$$\dot{\varphi}(t) = (\dot{\varphi}_1(t), \ldots, \dot{\varphi}_n(t)) = (f_1(t, \varphi(t)), \ldots, f_n(t, \varphi(t))).$$

Such a curve is often called an *integral curve.*

Given a point (t, x) in B, we may compute the value of the vector $f(t, x)$. If we construct a line segment $\varphi_{t,x}$ passing through (t, x) and parallel to $f(t, x)$, and do this for all (t, x) in B, we obtain the *direction field* of (3). It follows that any integral curve $\varphi(t)$, $r_1 < t < r_2$, of (3) is then tangent to $\varphi_{t,\varphi(t)}$ at each point $(t, \varphi(t))$ of B.

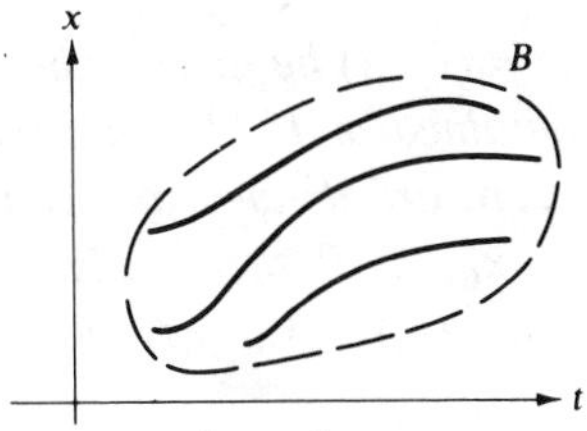

Integral curves

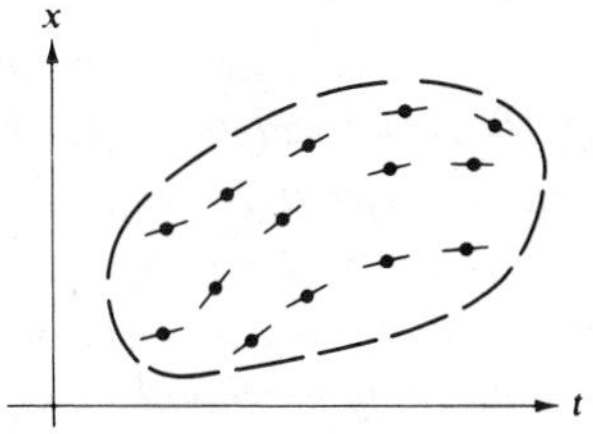

Direction field

Of particular interest will be the first-order system

$$\dot{x}_i = \sum_{j=1}^{n} a_{ij}(t)x_j + b_i(t), \qquad i = 1, \ldots, n,$$

where $a_{ij}(t)$, $i, j = 1, \ldots, n$, and $b_i(t)$, $i = 1, \ldots, n$, are continuous real-valued functions on $r_1 < t < r_2$ and $x(t) = (x_1(t), \ldots, x_n(t))$ is an unknown n-dimensional vector. If we denote by $A(t)$ the $n \times n$ matrix $(a_{ij}(t))$, and by $B(t)$ the vector $(b_1(t), \ldots, b_n(t))$, then the system can be conveniently expressed as

$$\dot{x} = A(t)x + B(t).$$

Hence $f(t, x) = A(t)x + B(t)$ and is defined in the infinite slab

$$B = \{(t, x) \mid r_1 < t < r_2, -\infty < x_i < \infty, i = 1, \ldots, n\}.$$

Observe that the form of equation (3) also suffices for the case of complex-valued unknown functions. For given the equation $\dot{z} = f(t, z)$, where $z = x + iy$, it can be written $\dot{x} + i\dot{y} = f(t, z) = h(t, x, y) + ig(t, x, y)$. This is equivalent to the system of equations $\dot{x} = h(t, x, y)$, $\dot{y} = g(t, x, y)$, where $x = x(t)$ and $y = y(t)$ are unknown functions, and is therefore in the form (3).

1.3 An Existence and Uniqueness Theorem

We now state a theorem giving sufficient conditions for the existence and uniqueness of solution of the first-order equation in normal form. The proof of the theorem is deferred until the last chapter inasmuch as the method of proof leads to several other results not relevant to our present discussion.

THEOREM 1.3.1. *Let the equation* $(*)$ $\dot{x} = f(t, x)$ *be given, where* $f(t, x)$ *is defined in some domain* B *contained in* R^{n+1}. *Suppose in addition that* f *and* $\partial f/\partial x_i$, $i = 1, \ldots, n$, *are defined and continuous in* B. *Then for every point* (t_0, x_0) *in* B *there exists a unique solution* $x = \varphi(t)$ *of* $(*)$ *satisfying* $\varphi(t_0) = x_0$ *and defined in some neighborhood of* (t_0, x_0).

Some remarks are in order. First of all, by *unique* is meant the following.

If two solutions $x = \varphi(t)$ and $x = \psi(t)$ of the equation $(*)$ both satisfy $\varphi(t_0) = \psi(t_0) = x_0$, then these solutions are identical in their common interval of definition.

Hence the theorem states that through every point of B there passes one and only one integral curve.

DEFINITION: The pair (t_0, x_0) are called the initial values of the solution $x = \varphi(t)$. The relation $x_0 = \varphi(t_0)$ is called the initial condition for the solution $\varphi(t)$.

The existence and uniqueness theorem is useful in the following sense: suppose by some technique we are able to find a family K of solutions of $\dot{x} = f(t, x)$. Furthermore, given a point (t_0, x_0) in B, there is an element φ in K satisfying $\varphi(t_0) = x_0$. If the hypotheses of

Theorem 1.3.1 are satisfied, then by uniqueness K must describe all solutions, and we need look no further for other solutions.

Example: Consider the equation

$$\dot{x} = f(t),$$

where $x = x(t)$ is an unknown scalar function, and $f(t)$ is continuous on $a < t < b$. The hypotheses of Theorem 1.3.1 are satisfied and B is the infinite strip $B = \{(t, x) \mid a < t < b, -\infty < x < \infty\}$. Consider the family K described by

$$\varphi(t; t_0, x_0) = x_0 + \int_{t_0}^{t} f(s) ds, \qquad a < t_0, t < b, \ -\infty < x_0 < \infty.$$

Any element of K satisfies $\dot{\varphi}(t; t_0, x_0) = f(t)$ and hence is a solution; given any (t_0, x_0) in B, then the element $\varphi(t; t_0, x_0)$ satisfies $\varphi(t_0; t_0, x_0) = x_0$. Therefore K describes all solutions of the equation.

Given any t_0 in (a, b), we let $F(t) = \int_{t_0}^{t} f(s)\, ds$, $a < t < b$, then any solution $\varphi(t)$ satisfying the initial conditions $\varphi(t_1) = x_1$ is given by $\varphi(t) = x_1 - F(t_1) + F(t)$, so all solutions are obtained by a translation of $F(t)$.

The integral curves look like this.

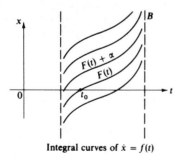

Integral curves of $\dot{x} = f(t)$

In the above example we were able to find an explicit form of the solution. The class of differential equations for which this can be done is small, but if we have a differential equation for which we know a solution exists, we may proceed to investigate properties of the solution (such as behavior for large t or boundedness), regardless of whether we know its explicit form. The qualitative study of solutions of differential equations thus depends upon existence.

Uniqueness would be of importance if, for instance, we wished to approximate the solution numerically. If two solutions passed through a point, then successive approximations could very well jump from one solution to the other—with misleading consequences. Furthermore, uniqueness assures us that the domain B is smoothly covered by a nonintersecting family of integral curves. Such a fact is useful in the study of dynamic or spatial properties of solutions.

It should be remarked that if in the hypotheses of Theorem 1.3.1 we only require that f be continuous in B, then existence of a solution satisfying $\varphi(t_0) = x_0$ is still guaranteed. However, uniqueness may no longer occur.

Example: The functions $\varphi(t) = (t - t_0)^3$ and $\psi(t) = 0$, $-\infty < t_0, t < \infty$, are easily seen to be solutions of the equation $\dot{x} = 3x^{2/3}$. Hence, given the initial values $(t_0, 0)$, $-\infty < t_0 < \infty$, there are two solutions satisfying them. Note that $f(x) = 3x^{2/3}$ is continuous at $x = 0$, but $df/dx = 2x^{-1/3}$ fails to exist there. For the domains $B = \{(t, x) | -\infty < t < \infty, x > 0\}$ or $B = \{(t, x) | -\infty < t < \infty, x < 0\}$, both f and df/dx are continuous, and the unique solution satisfying $\varphi(t_0) = x_0$ for $x_0 \neq 0$ is given by $\varphi(t) = (t - t_0 + x_0^{1/3})^3$. It is defined for $-\infty < t < \infty$, but fails to be unique at $t = t_0 - x_0^{1/3}$, where it intersects the zero solution.

1.4 The Maximum Interval of Existence

Suppose we are given two solutions $\varphi_1(t)$, $r_1 < t < r_2$, and $\varphi_2(t)$, $s_1 < t < s_2$, of the differential equation $\dot{x} = f(t, x)$, and both solutions satisfy the initial condition. Therefore $\varphi_1(t_0) = \varphi_2(t_0) = x_0$. If the equation satisfies the hypotheses of Theorem 1.3.1, then in a neighborhood of (t_0, x_0) we have uniqueness, and the two solutions overlap. That is, if for instance $r_1 < s_1 < t_0 < r_2 < s_2$, then $\varphi_1(t) = \varphi_2(t)$ for $s_1 < t < r_2$.

However, we can define a new solution $\varphi(t)$ defined on $r_1 < t < s_2$, and containing both $\varphi_1(t)$ and $\varphi_2(t)$, as follows:

$$\varphi(t) = \varphi_1(t) \qquad \text{if} \qquad r_1 < t < r_2,$$

$$\varphi(t) = \varphi_2(t) \qquad \text{if} \qquad s_1 < t < s_2.$$

It is a solution, since $\varphi_1(t)$ and $\varphi_2(t)$ are solutions, it agrees with their common values on $s_1 < t < r_2$, and it is defined on the larger interval $r_1 < t < s_2$. This same procedure of tacking together solutions would apply if we were given a finite number of solutions $\varphi_1(t), \ldots, \varphi_m(t)$, such that $\varphi_1(t_0) = \cdots = \varphi_m(t_0) = x_0$. We could then define a new solution $\varphi(t)$ satisfying $\varphi(t_0) = x_0$ and whose interval of definition contains those of $\varphi_1(t), \ldots, \varphi_m(t)$.

We might conjecture that, given initial values (t_0, x_0), there exists a solution $\varphi(t)$, $m_1 < t < m_2$, whose interval of definition (m_1, m_2) is maximal in some sense. The following theorem indicates that this is true.

THEOREM 1.4.1. *Suppose the hypotheses of Theorem* 1.3.1 *are satisfied for the differential equation* (*) $\dot{x} = f(t, x)$. *Then, given initial values* (t_0, x_0), *there exists a solution* $\varphi(t)$ *of* (*), *defined on* $m_1 < t < m_2$, *satisfying* $\varphi(t_0) = x_0$; *furthermore, if* $\psi(t)$ *is any other solution and* $\psi(t_0) = x_0$, *then its interval of definition is contained in* (m_1, m_2).

Proof: Let M be the set of all intervals of definition of solutions of (*) satisfying the initial values, and M is not empty, since at least one such solution exists. Let M_1 be the set of all left end points of the elements of M and M_2 the set of all right end points. Let $m_1 = \inf M_1$ and $m_2 = \sup M_2$ and suppose t_1 belongs to (m_1, m_2). Then there exists a solution $\psi(t)$ whose interval of definition contains t_1 and we define $\varphi(t_1) = \psi(t_1)$. By uniqueness, $\varphi(t)$ is well-defined on $m_1 < t < m_2$ and is a solution, since for every t it agrees with a solution, and finally $\varphi(t_0) = x_0$. By the construction, (m_1, m_2) is maximal in the sense described.

DEFINITION: The interval (m_1, m_2) is called the maximum interval of existence corresponding to the initial values (t_0, x_0).

Examples

(*a*) If we consider the equation $\dot{x} = f(t)$ previously discussed in Section 1.3, and $f(t)$ is continuous on $a < t < b$, then, given initial values (t_0, x_0), the corresponding solution is

$$\varphi(t) = x_0 + \int_{t_0}^{t} f(s)\, ds.$$

The maximum interval of existence is (a, b), since the integral is by hypothesis only defined for $a < t < b$. For instance, if $f(t) = e^t$, then $(a, b) = (-\infty, \infty)$, whereas if

$$f(t) = \frac{1}{t(1-t)},$$

then (a, b) can be $(0, 1)$, $(1, \infty)$ or $(-\infty, 0)$, depending on the choice of t_0.

(*b*) This example shows that the maximum interval of existence may vary considerably with the choice of initial values. Given the equation $\dot{x} = -3x^{4/3} \sin t$, $x(t)$ an unknown scalar function, its solutions are $x(t) = 0$ and $x(t) = (c - \cos t)^{-3}$, where the constant c is determined by the choice of initial values (t_0, x_0). The hypotheses of Theorem 1.3.1 are satisfied where B is the whole (t, x) plane.

The solution $x(t) = 0$ is defined on $-\infty < t < \infty$, and hence any initial value $(t_0, 0)$ has $(-\infty, \infty)$ as its maximum interval of existence. The second solution is defined on $-\infty < t < \infty$ if $|c| > 1$, whereas it will only be defined on a finite interval if $|c| \leq 1$. For example, the solution $x(t)$ satisfying $x(\pi/2) = 1/8$ is

$$x(t) = (2 - \cos t)^{-3}, \qquad -\infty < t < \infty,$$

whereas the solution satisfying $x(\pi/2) = 8$ is

$$x(t) = (1/2 - \cos t)^{-3}, \qquad \pi/3 < t < 5\pi/3,$$

and both the intervals are maximal for their respective initial values.

A sketch of the solutions with initial values $(\pi/2, 0)$, $(\pi/2, 1/8)$ and $(\pi/2, 8)$ would look like the figure on page 13.

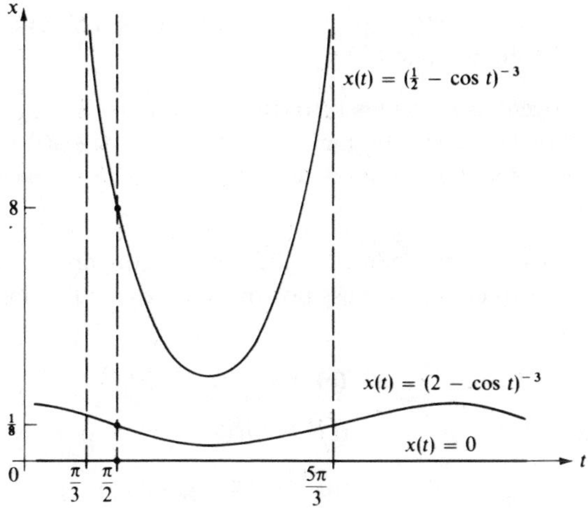

Solutions of $\dot{x} = -3x^{4/3} \sin t$

Problems

1. Write the following equations as a first-order system in normal form.
(a) $x^{(4)} + x^{(3)} \cos t - \ddot{x} + x^2 \sin t = 0$, $x = x(t)$ a scalar function.
(b) $\ddot{u} + u = \dot{v} \sin t$, $\ddot{v} + v = \dot{u} \cos t$, $u = u(t)$, $v = v(t)$ scalar functions.

2. Given the equation $\dot{x} = f(x)$, where $x = x(t)$ is an unknown scalar function, and f and df/dx are defined and continuous on the strip $B = \{(t, x) \mid -\infty < t < \infty, a < x < b\}$. Assume $f(x) \neq 0$, $a < x < b$, and let

$$F(x) = \int_{x_0}^{x} \frac{1}{f(s)}\, ds,$$

where $a < x_0 < b$. Show all solutions are described by the family $x(t) = \varphi(t - c)$, where φ is the inverse of F, and the constant c is determined by the initial conditions.

3. Given the equation $\dot{x} = f(t)g(x) \equiv F(t, x)$, where $x = x(t)$ is an unknown scalar function, and F and $\partial F/\partial x$ are defined and continuous on the rectangle $B = \{(t, x) \mid t_1 < t < t_2, a < x < b\}$. Assume that $g(x) \neq 0$, $a < x < b$.

Let $u = \varphi(t)$ be the solution of $\dot{u} = f(t)$ and $x = \psi(u)$ be the solution of $dx/du = g(x)$. Show that all solutions of $\dot{x} = F(t, x)$ are described by the family $x(t) = \psi(\varphi(t) - c)$, where the constant c is determined by the initial conditions.

4. If $f(t, x)$ is homogenous of degree zero (that is, $f(\alpha t, \alpha x) = f(t, x)$ for $\alpha \neq 0$), show that the substitution $x = tu$ or $x = t/u$, $x(t)$, $u(t)$ scalar functions, transforms the equation $\dot{x} = f(t, x)$ into the type considered in Problem 3.

5. Find all solutions of the following differential equations, where $x = x(t)$ is a scalar function. Discuss possible choices for the domain B of Theorem 1.3.1.

(a) $\dot{x} = te^t$.

(b) $\dot{x} = t \log(t^2 - 1)$.

(c) $\dot{x} = x^2 - 4$.

(d) $\dot{x} = \sec x$.

(e) $\dot{x} = -(t + 1)x/t$.

(f) $\dot{x} = t^3(x + 1)^{-2}$.

(g) $\dot{x} = (x + t)/t$.

(h) $\dot{x} = (x - \sqrt{x^2 + t^2})/t$.

6. Describe all solutions of $\dot{x} = (x^2 - 1)/xt$, $x = x(t)$ a scalar function, and find the maximum interval of existence of the solution satisfying $x(1) = \sqrt{2}/2$.

7. Solutions need not be given as explicit functions of t. For example, show that a solution of $\dot{x} = (t + 1)(x + 1)/tx$, $x = x(t)$ a scalar function, is given by $t(x + 1) = 2e^{x-t}$.

8. Given the Riccati equation $\dot{x} = p(t)x + q(t)x^2 + r(t)$, where $x = x(t)$ is a scalar function and p, q, and r are continuous on $a < t < b$. Show that if $x = \varphi(t)$ is a solution, then there are further solutions of the form $x = \varphi(t) + 1/\psi(t)$, where $\psi(t)$ is a solution of an equation of the form $\dot{x} = c(t)x + d(t)$.

CHAPTER

2

The Linear Equation: General Discussion

2.1 Introduction

In this chapter we will discuss the n-dimensional first-order system of equations

$$\dot{x}_i = \sum_{j=1}^{n} a_{ij}(t)x_j + b_i(t), \qquad i = 1, \ldots, n, \tag{1}$$

where $x(t) = (x_1(t), \ldots, x_n(t))$ is an unknown vector function, and $a_{ij}(t)$ and $b_i(t)$, $i, j = 1, \ldots, n$, are given continuous functions on $r_1 < t < r_2$.

As remarked in Section 1.2, this equation can be conveniently written as

$$\dot{x} = A(t)x + B(t), \tag{2}$$

where $A(t)$ is the $n \times n$ matrix $(a_{ij}(t))$ and $B(t)$ is the vector $(b_1(t), \ldots, b_n(t))$. The function $f(t, x) = A(t)x + B(t)$ satisfies the hypotheses of Theorem 1.3.1 in the domain

$$B = \{(t, x) \mid r_1 < t < r_2, -\infty < x_i < \infty, i = 1, \ldots, n\}.$$

Therefore, given initial values in B, there exists a unique solution of (2) satisfying them.

Examples

(*a*) A number of physical processes involving growth or decay are described by the differential equation

$$\dot{x} = a(t)x.$$

Here $x = x(t)$ is an unknown scalar function representing, for example, the population or amount of material remaining at time t, and $a(t)$ is a continuous funtion on $0 < t < \infty$ and represents a growth or decay factor.

The equation is of the form (2), with $A(t) = a(t)$ and $B(t) = 0$, and the solution (see Problem 3, Chapter 1) satisfying $x(t_0) = x_0$ is given by

$$x(t) = x_0 \exp\left[\int_{t_0}^{t} a(s)\, ds\right].$$

(*b*) The nth-order equation

$$y^{(n)} + a_1(t)y^{(n-1)} + \cdots + a_n(t)y = b(t),$$

where $y = y(t)$ is an unknown scalar function and $a_i(t)$, $i = 1, \ldots, n$, and $b(t)$ are continuous functions on $r_1 < t < r_2$, is of extreme importance both mathematically and in physical applications.

By the substitution (see Section 1.2) $x_1 = y$, $x_2 = \dot{y}, \ldots,$ $x_n = y^{(n-1)}$, the equation can be written in the form (2), where

$$A(t) = \begin{pmatrix} 0 & 1 & 0 & \cdots & & 0 \\ 0 & 0 & 1 & \cdots & & 0 \\ \vdots & & & \vdots & & \vdots \\ 0 & & & \cdots & 0 & 1 \\ -a_n(t) & & & \cdots & & -a_1(t) \end{pmatrix}, \qquad B(t) = \begin{pmatrix} 0 \\ \vdots \\ 0 \\ b(t) \end{pmatrix}.$$

The following theorem gives an important property of solutions of (2). Its proof is deferred until the last chapter, since it is a consequence of the proof of the existence and uniqueness theorem.

THEOREM 2.1.1. *Given the equation* $\dot{x} = A(t) + B(t)$, *where* $A(t)$ *and* $B(t)$ *are continuous on* $r_1 < t < r_2$, *then for any initial value* (t_0, x_0), $r_1 < t_0 < r_2$, *there exists a solution defined on* $r_1 < t < r_2$ *and satisfying the initial value.*

The theorem states that the maximal interval of continuity of $A(t)$ and $B(t)$ is the maximum interval of existence for any initial value. Thus, for instance, if $a_i(t)$, $i = 1, \ldots, n$, and $b(t)$ are continous for all t, then any solution $y(t)$ of the equation given in Example (b) can be defined for $-\infty < t < \infty$.

2.2 Fundamental Solutions

We will first discuss the first-order linear homogeneous system given by

$$\dot{x} = A(t)x, \tag{3}$$

where $x = x(t) = (x_1(t), \ldots, x_n(t))$ is an unknown n-dimensional vector function, and $A(t) = (a_{ij}(t))$ is an $n \times n$ matrix that is continuous on $r_1 < t < r_2$. From the previous discussion it follows that given (t_0, x_0), $r_1 < t_0 < r_2$, there exists a unique solution $x(t)$, defined on $r_1 < t < r_2$ and satisfying $x(t_0) = x_0$.

The system (3) is of the form discussed in the previous section and is called *homogeneous*, since $B(t) = 0$. It is called *linear*, since any linear combination of solutions of (3) is also a solution. Specifically, let $\varphi_i(t) = (\varphi_{1i}(t), \ldots, \varphi_{ni}(t))$, $i = 1, \ldots, m$, be solutions of (3), and let c_i, $i = 1, \ldots, m$, be arbitrary constants. Then, for $\varphi(t) = \sum_{i=1}^m c_i \varphi_i(t)$, we have

$$\dot{\varphi} = \sum_{i=1}^{m} c_i \dot{\varphi}_i = \sum_{i=1}^{m} c_i(A(t)\varphi_i) = A(t)\left(\sum_{i=1}^{m} c_i \varphi_i\right) = A(t)\varphi$$

by the properties of matrix-vector multiplication. But the relation $\dot{\varphi} = A(t)\varphi$ implies that $\varphi(t)$ is also a solution of (3).

Note that $\varphi(t) \equiv 0$, $r_1 < t < r_2$, is a solution of (3), and in fact is the only solution satisfying $\varphi(t_0) = 0$ for $r_1 < t_0 < r_2$, as the following lemma shows.

LEMMA 2.2.1. *If* $r_1 < t_0 < r_2$ *and* $x = \varphi(t)$ *is a solution of* (3) *satisfying* $\varphi(t_0) = 0$, *then* $\varphi(t)$ *is identically zero on* $r_1 < t < r_2$.

Proof. The solution $x(t) \equiv 0, r_1 < t < r_2$, satisfies $x(t_0) = 0$, but by uniqueness we must have $x(t) = \varphi(t)$, since all solutions are defined on $r_1 < t < r_2$.

We now introduce the key notion of linear independence of a given collection of scalar or vector-valued functions.

DEFINITION: A collection of functions $\alpha_1(t), \ldots, \alpha_m(t)$, $a < t < b$, is linearly independent if there exist no constants $c_1, \ldots, c_m$, not all zero, such that $\sum_{j=1}^{m} c_j \alpha_j(t)$ is identically zero on (a, b). The collection is linearly dependent if it is not linearly independent.

Examples

(*a*) The collection of scalar functions $1, t, t^2, \ldots, t^m$, $-\infty < t < \infty$, is linearly independent, since $P_m(t) = \sum_{j=0}^{m} c_j t^j \equiv 0$, $-\infty < t < \infty$, can only occur when $c_0, c_1, \ldots, c_m$ are all zero, since a non-trivial polynomial can only have a finite number of zeros.

(*b*) The collection of n-dimensional vector functions

$$\alpha_j(t) = (0, \ldots, 0, \overset{j\text{th place}}{1}, 0, \ldots, 0) = e_j,$$

$j = 1, \ldots, n$, $-\infty < t < \infty$, is linearly independent, since $\sum_{j=1}^{n} c_j \alpha_j(t) = (c_1, \ldots, c_n) = 0$ if and only if $c_j = 0$ for $j = 1, \ldots, n$.

(*c*) The vector functions $\alpha_1(t) = (\sqrt{2}\, t, \cos t)$, $\alpha_2(t) = (\sqrt{2}\, t^2, \sin t)$ and $\alpha_3(t) = (t - t^2, \cos(t + \pi/4))$ are linearly dependent on $-\infty < t < \infty$, since $\sum_{j=1}^{3} c_j \alpha_j(t) \equiv 0$, $-\infty < t < \infty$, for $c_1 = \sqrt{2}/2$, $c_2 = -\sqrt{2}/2$, and $c_3 = -1$.

LEMMA 2.2.2. *If* $\varphi_1(t), \ldots, \varphi_m(t)$, $r_1 < t < r_2$, *are a collection of linearly independent solutions of* (3), *then the linear combination* $\sum_{j=1}^{m} c_j \varphi_j(t)$ *never vanishes on* $r_1 < t < r_2$ *unless* $c_1 = \cdots = c_m = 0$.

Proof. If we let $\varphi(t) = \sum_{j=1}^{m} c_j \varphi_j(t)$, then, by linearity, $\varphi(t)$ is a solution of (3). If $\varphi(t_0) = 0$ for some t_0 in (r_1, r_2) and the c_j are not all zero, then by Lemma 2.2.1 we conclude that $\varphi(t)$ is identically zero, which contradicts linear independence.

DEFINITION: A collection $\varphi_1(t), \ldots, \varphi_n(t)$, $r_1 < t < r_2$, of solutions of the n-dimensional first-order linear system (3) is called a *fundamental system of solutions* of (3) if it is linearly independent.

The importance of a fundamental system of solutions of a linear system is that we may describe any solution of the system in terms of the fundamental system of solutions. The problem of finding any solution then becomes one of finding n linearly independent solutions; but even more important, we need only properties of the fundamental system in order to determine the behavior of any solution.

We now show that a fundamental system of solutions exists for equation (3).

THEOREM 2.2.1. *A fundamental system of solutions of* (3) *exists.*

Proof. Let $e_1, \ldots, e_n$ be the linearly independent set of n-dimensional vectors

$$e_j = (0, \ldots, 0, \overset{j\text{th place}}{1}, 0, \ldots, 0), \qquad j = 1, \ldots, n.$$

(See Example (*b*) above.) For any t_0 in (r_1, r_2), let $\varphi_1(t), \ldots, \varphi_n(t)$ be the solutions of (3) satisfying $\varphi_j(t_0) = e_j$, $j = 1, \ldots, n$. These are all distinct, since each satisfies distinct initial values. Furthermore, they are linearly independent, for if $\varphi(t) = \sum_{j=1}^n c_j \varphi_j(t) \equiv 0, r_1 < t < r_2$, with the c_j not all zero, then

$$\varphi(t_0) = \sum_{j=1}^{n} c_j \varphi_j(t_0) = \sum_{j=1}^{n} c_j e_j = (c_1, \ldots, c_n) = 0.$$

But this implies $c_1 = \cdots = c_n = 0$, which is a contradiction. Therefore $\varphi_1(t), \ldots, \varphi_n(t)$ are linearly independent, and form a fundamental system of solutions of (3).

COROLLARY 1. *Every solution of equation* (3) *is a linear combination of the elements of a fundamental system of solutions.*

Proof. If $x(t)$ is a solution of (3) it is defined for $r_1 < t < r_2$; let $x_0 = (x_{10}, \ldots, x_{n0})$ be its value at t_0. Let $\varphi_1(t), \ldots, \varphi_n(t)$ be the fundamental system of solutions constructed above and let $\varphi(t) = \sum_{j=1}^{n} x_{j0}\, \varphi_j(t)$. By linearity, $\varphi(t)$ is a solution, and furthermore $\varphi(t_0) = (x_{10}, \ldots, x_{n0}) = x_0$. By uniqueness, we must have $\varphi(t) = x(t)$ for $r_1 < t < r_2$.

Examples

(*a*) For the equation $\dot{x} = a(t)x$, with $a(t)$ a scalar continuous function on $r_1 < t < r_2$, let $\varphi(t) = \exp\left[\int_{t_0}^{t} a(t)\,dt\right]$. It is a solution, it is never zero on $r_1 < t < r_2$, and hence it is trivially independent and satisfies $\varphi(t_0) = 1$. Any solution $x(t)$ is therefore given by $x(t) = x_0\, \varphi(t)$, $r_1 < t < r_2$, and $x(t_0) = x_0$.

(*b*) Consider the two-dimensional linear system $\dot{x} = A(t)x$ corresponding to the second-order equation $\ddot{x} + x = 0$, where $x(t)$ is an unknown scalar function. This system is given by

$$\dot{x} = y, \qquad \dot{y} = -x,$$

and hence

$$A(t) = \begin{pmatrix} 0 & 1 \\ -1 & 0 \end{pmatrix},$$

and all solutions $(x(t), y(t))$ are defined on $-\infty < t < \infty$. Consider the two solutions $\varphi_1(t) = (\cos t, -\sin t)$ and $\varphi_2(t) = (\sin t, \cos t)$. That they are linearly independent follows from the fact that $c_1 \sin t + c_2 \cos t \equiv 0$ implies $c_1 = c_2 = 0$, (let $t = 0$; then $t = \pi/2$). Furthermore, $\varphi_1(0) = (1, 0)$, $\varphi_2(0) = (0, 1)$, so any solution $(x(t), y(t))$ is given by $(x(t), y(t)) = x_0\, \varphi_1(t) + y_0\, \varphi_2(t)$ for $-\infty < t < \infty$, where $(x(0), y(0)) = (x_0, y_0)$. This solution corresponds to the solution $x = \varphi(t)$ of the equation $\ddot{x} + x = 0$ satisfying the initial conditions $\varphi(0) = x_0$, $\dot{\varphi}(0) = y_0$.

The discussion and the theorem indicate the following: the space X of all solutions of (3) is a linear space, since any linear combination of solutions is again a solution. Furthermore, X has dimension n, since any element of the space can be described as a linear combination of the n linearly independent elements of a fundamental system of solutions.

Finally, we introduce the $n \times n$ matrix $\Phi(t)$, whose jth column is $\varphi_j(t)$ as constructed in Theorem 2.2.1. If we let $\varphi_j(t) = (\varphi_{1j}(t), \ldots, \varphi_{nj}(t))$, $j = 1, \ldots, n$, then $\Phi(t) = (\varphi_{ij}(t))$ and $\Phi(t_0) = I$, the identity matrix. By $\dot{\Phi}(t)$ we mean the matrix $(\dot{\varphi}_{ij}(t))$, and from the above we have

COROLLARY 2. *The matrix $\Phi(t)$ is the solution of the matrix differential equation $\dot{\Phi} = A(t)\Phi$ satisfying $\Phi(t_0) = I$. Furthermore, the solution $x(t)$ of (3) satisfying $x(t_0) = x_0$ can be written $x(t) = \Phi(t)x_0$.*

Proof. Each column of $\Phi(t)$ is a solution $\varphi_j(t)$ of (3) and hence satisfies $\dot{\varphi}_j = A(t)\varphi_j$; therefore $\Phi(t)$ satisfies $\dot{\Phi} = A(t)\Phi$. If $x_0 = (x_{10}, \ldots, x_{n0})$, then by Corollary 1 we have

$$x(t) = \sum_{j=1}^{n} x_{j0}\, \varphi_j(t) = \Phi(t)x_0 .$$

2.3 The Wronskian

We continue our discussion of the n-dimensional linear system

$$\dot{x} = A(t)x, \tag{4}$$

and will specifically determine a necessary and sufficient condition for a collection $\varphi_1(t), \ldots, \varphi_n(t)$ of solutions of (4) to be a fundamental system of solutions—that is, to be linearly independent. As in previous sections we assume that $A(t)$ is continuous for $r_1 < t < r_2$.

DEFINITION: Let $\varphi_1(t), \ldots, \varphi_n(t)$ be solutions of (4) where $\varphi_j(t) = (\varphi_{1j}(t), \ldots, \varphi_{nj}(t))$. Then the scalar function

$$W(t) = \det\begin{pmatrix} \varphi_{11}(t) & \cdots & \varphi_{1n}(t) \\ \vdots & & \vdots \\ \varphi_{n1}(t) & \cdots & \varphi_{nn}(t) \end{pmatrix}$$

is called the *Wronskian* of $\varphi_1(t), \ldots, \varphi_n(t)$.

DEFINITION: If $\varphi_1(t), \ldots, \varphi_n(t)$ is a fundamental system of solutions of (4), then the matrix corresponding to $W(t)$ is called a *fundamental matrix.*

Thus a fundamental matrix is a matrix whose columns form a fundamental system of solutions of (4). The matrix $\Phi(t)$ constructed in the previous section is a fundamental matrix.

Example: For the two-dimensional system $\dot{x} = y$, $\dot{y} = x$, given in the previous section, the matrix

$$\Phi(t) = \begin{pmatrix} \cos t & \sin t \\ -\sin t & \cos t \end{pmatrix}, \qquad -\infty < t < \infty,$$

is a fundamental matrix, and $W(t) \equiv 1$, $-\infty < t < \infty$.

The following theorem shows that given any n solutions of (4) and any t_0 in (r_1, r_2), we can completely determine the corresponding Wronskian without computing the $n \times n$ determinant. Recall that for any square matrix $A = (a_{ij})$, the trace of A is given by

$$\operatorname{tr} A = \sum_{i=1}^{n} a_{ii}.$$

THEOREM 2.3.1. (*Liouville's formula.*) Let $\varphi_1(t), \ldots, \varphi_n(t)$ *be any n solutions of equation* (4) *and let t_0 be in (r_1, r_2). Then the Wronskian of $\varphi_1(t), \ldots, \varphi_n(t)$ is given by*

$$W(t) = W(t_0) \exp\left[\int_{t_0}^{t} \operatorname{tr} A(s)\, ds\right], \qquad r_1 < t < r_2.$$

Proof. We will show that $W(t)$ satisfies the differential equation $\dot{W} = [\operatorname{tr} A(t)]W$, from which the conclusion follows (see Example (*a*), Section 2.1). If $\varphi_j(t) = (\varphi_{1j}(t), \ldots, \varphi_{nj}(t))$, then by the expansion by cofactors of $W(t)$ we have

$$W(t) = \sum_{j=1}^{n} \varphi_{ij}(t)\,|W_{ij}(t)|,$$

where $|W_{ij}(t)|$ is the i, jth cofactor of $W(t)$.

The cofactor $|W_{ij}(t)|$ does not contain the term $\varphi_{ij}(t)$, so if we regard W as a function of the φ_{ij} we have $\partial W/\partial\varphi_{ij} = |W_{ij}|$ and, by the chain rule,

$$\dot{W} = \sum_{i,j=1}^{n} \frac{\partial W}{\partial \varphi_{ij}}\,\dot{\varphi}_{ij} = \sum_{i=1}^{n}\left(\sum_{j=1}^{n} \dot{\varphi}_{ij}\,|W_{ij}|\right).$$

Let $\mathscr{W}$ be the matrix corresponding to $W(t)$, differentiate its ith row, and call the corresponding determinant $W_i(t)$. The expression in parentheses in the above equation is the expansion of $W_i(t)$ by cofactors, and therefore $\dot{W}(t) = \sum_{i=1}^n W_i(t)$.

Each column $(\varphi_{1j}(t), \ldots, \varphi_{nj}(t))$ of $\mathscr{W}$ satisfies (4) and therefore $\dot{\varphi}_{ij} = \sum_{k=1}^n a_{ik}(t)\varphi_{kj}$ for $i = 1, \ldots, n$) hence

$$W_i(t) = \det\begin{pmatrix} \varphi_{11} & \cdots & \varphi_{1n} \\ \vdots & & \vdots \\ \dot{\varphi}_{i1} & \cdots & \dot{\varphi}_{in} \\ \vdots & & \vdots \\ \varphi_{n1} & \cdots & \varphi_{nn} \end{pmatrix} = \det\begin{pmatrix} \varphi_{11} & \cdots & \varphi_{1n} \\ \vdots & & \vdots \\ \sum_{k=1}^n a_{ik}\,\varphi_{k1} & \cdots & \sum_{k=1}^n a_{ik}\,\varphi_{kn} \\ \vdots & & \vdots \\ \varphi_{n1} & \cdots & \varphi_{nn} \end{pmatrix}.$$

Multiply the kth row for $k \neq i$ of the last matrix given by $-a_{ik}(t)$ and then add it to the ith row. This does not change the value of the determinant but gives the relation

$$W_i(t) = \det\begin{pmatrix} \varphi_{11} & \cdots & \varphi_{1n} \\ \vdots & & \vdots \\ a_{ii}\,\varphi_{i1} & \cdots & a_{ii}\,\varphi_{in} \\ \vdots & & \vdots \\ \varphi_{n1} & \cdots & \varphi_{nn} \end{pmatrix} = a_{ii}(t)W(t).$$

It follows that $\dot{W}(t) = \sum_{i=1}^n a_{ii}(t)W(t) = [\operatorname{tr} A(t)]W(t)$, which leads to the desired result.

Since $\exp\left[\int_{t_0}^t \operatorname{tr} A(s)\,ds\right]$ is never zero, the theorem implies that the Wronskian of any collection of n solutions of (4) is identically zero—$W(t_0) = 0$ for some t_0—or never zero on (r_1, r_2). The latter case characterizes a fundamental system as the following theorem shows.

THEOREM 2.3.2. *A necessary and sufficient condition for $\varphi_1(t), \ldots, \varphi_n(t)$ to be a fundamental system of solutions of* (4) *is that $W(t) \neq 0$ for $r_1 < t < r_2$.*

Proof. Let $\varphi_1(t), \ldots, \varphi_n(t)$ be a fundamental system of solutions of (4) and let $\varphi(t)$ be any nontrivial solution. By Corollary 1 of Theorem 2.2.2 there exist $c_1, \ldots, c_n$ not all zero such that $\varphi(t) = \sum_{i=1}^{n} c_i \varphi_i(t)$, and by the uniqueness of the solutions the c_i are unique. If $c = (c_1, \ldots, c_n)$ and $\Phi(t)$ is the fundamental matrix of $\varphi_1(t), \ldots, \varphi_n(t)$, then the previous relation can be written $\varphi(t) = \Phi(t)c$. For any t in (r_1, r_2) this is a system of n linear equations in the unknowns $c_1, \ldots, c_n$, which has a unique solution, and this implies $\det \Phi(t) = W(t) \neq 0$.

Conversely, if $W(t) \neq 0$ for $r_1 < t < r_2$, this implies that the columns $\varphi_1(t), \ldots, \varphi_n(t)$ of $\Phi(t)$ are linearly independent for $r_1 < t < r_2$. Since they are solutions of (4), they form a fundamental system of solutions.

Finally, equipped with the above two results we are able to state the following sharper version of Corollary 2 of Theorem 2.2.1.

COROLLARY. *Any fundamental matrix $\Phi(t)$ is the solution of the matrix differential equation $\dot{\Phi} = A(t)\Phi$. Furthermore, the solution $x(t)$ of (4) satisfying $x(t_0) = x_0$ can be written $x(t) = \Phi(t)\Phi^{-1}(t_0)x_0$, and the matrix $\Omega(t) = \Phi(t)\Phi^{-1}(t_0)$ is a fundamental matrix satisfying $\Omega(t_0) = I$.*

Proof. We have left only to prove the second statement. Given the fundamental matrix $\Phi(t)$, the solution $x(t)$ satisfies $x(t) = \Phi(t)c$ for some constant vector c, and therefore $x_0 = \Phi(t_0)c$. But $W(t_0) = \det \Phi(t_0) \neq 0$ implies that $\Phi^{-1}(t_0)$ exists, and thus $c = \Phi^{-1}(t_0)x_0$, which gives the first result.

Furthermore, $\det \Phi^{-1}(t_0) = W^{-1}(t_0) \neq 0$, and this implies that the columns of $\Omega(t) = \Phi(t)\Phi^{-1}(t_0)$ are linearly independent, since those of $\Phi(t)$ were. The columns of $\Omega(t)$ are linear combinations of those of $\Phi(t)$ and are therefore solutions of (4). Hence $\Omega(t)$ is a fundamental matrix and $\Omega(t_0) = I$.

Example: Consider the two-dimensional system

$$\dot{x} = 2x + y, \qquad \dot{y} = 3x + 4y,$$

and hence

$$A(t) = \begin{pmatrix} 2 & 1 \\ 3 & 4 \end{pmatrix},$$

and $\operatorname{tr} A(t) = 2 + 4 = 6$. All solutions are defined on $-\infty < t < \infty$, so if we let $t_0 = 0$ the Wronskian of any fundamental system of solutions is

$$W(t) = W(0) \exp\left[\int_0^t 6\, ds\right] = W(0)e^{6t}.$$

A fundamental system of solutions is given by

$$\varphi_1(t) = (x(t), y(t)) = (e^t, -e^t)$$

and

$$\varphi_2(t) = (e^{5t}, 3e^{5t}).$$

Therefore

$$\Phi(t) = \begin{pmatrix} e^t & e^{5t} \\ -e^t & 3e^{5t} \end{pmatrix},$$

and hence $W(0) = \det \Phi(0) = 4$. The Wronskian is therefore $W(t) = 4e^{6t}$, as may be verified by direct calculation.

Finally, we have

$$\Phi(0) = \begin{pmatrix} 1 & 1 \\ -1 & 3 \end{pmatrix}, \qquad \Phi^{-1}(0) = \begin{pmatrix} \frac{3}{4} & -\frac{1}{4} \\ \frac{1}{4} & \frac{1}{4} \end{pmatrix},$$

and hence

$$\Omega(t) = \Phi(t)\Phi^{-1}(0) = \begin{pmatrix} \frac{3}{4}e^t + \frac{1}{4}e^{5t} & -\frac{1}{4}e^t + \frac{1}{4}e^{5t} \\ -\frac{3}{4}e^t + \frac{3}{4}e^{5t} & \frac{1}{4}e^t + \frac{3}{4}e^{5t} \end{pmatrix}$$

is a fundamental matrix satisfying $\Omega(0) = I$. Therefore any solution $\varphi(t)$ can be given by

$$\varphi(t) = \Omega(t)\varphi(0).$$

2.4 The Nonhomogeneous Linear Equation

In this section we introduce the method of *variation of parameters* and use it to obtain the solution of the first-order n-dimensional nonhomogeneous equation

$$\dot{x} = A(t)x + B(t). \tag{5}$$

Here, as before, $A(t) = (a_{ij}(t))$, $i, j = 1, \ldots, n$, and $B(t) = (b_1(t) \ldots, b_n(t))$, are assumed to be continuous on $r_1 < t < r_2$. By our previous discussion this implies that solutions exist and are unique in

$$B = \{(t, x) \mid r_1 < t < r_2, \ -\infty < x_i < \infty, \ i = 1, \ldots, n\},$$

and every such solution is defined on $r_1 < t < r_2$.

In the previous section we obtained the result that the solution $x(t)$ of $\dot{x} = A(t)x$ satisfying $x(t_0) = x_0$ is given by $x(t) = \Phi(t)x_0$, where $\Phi(t)$ is a fundamental matrix satisfying $\Phi(t_0) = I$. To apply the method of variation of parameters, we assume the solution $x(t)$ of (5) satisfying $x(t_0) = x_0$ can be given by $x(t) = \Phi(t)c(t)$, where $c(t) = (c_1(t), \ldots, c_n(t))$. This leads to

THEOREM 2.4.1. *The solution $x(t)$ of equation* (5) *satisfying $x(t_0) = x_0$, $r_1 < t_0 < r_2$, is given by*

$$x(t) = \Phi(t)x_0 + \int_{t_0}^{t} \Phi(t)\Phi^{-1}(s)B(s)\,ds, \qquad r_1 < t < r_2,$$

where $\Phi(t)$ is the fundamental matrix of the equation $\dot{x} = A(t)x$ satisfying $\Phi(t_0) = I$.

Proof. The representation $x(t) = \Phi(t)c(t)$, $c(t) = (c_1(t), \ldots, c_n(t))$ is valid if and only if (i) $c(t_0) = x_0$, since $\Phi(t_0) = I$, and (ii) $\Phi(t)\dot{c} = B(t)$. This last result is obtained by substituting $x(t) = \Phi(t)c(t)$ into (5). By the product rule for differentiation this gives

$$\dot{x} = \dot{\Phi}c + \Phi\dot{c} = A(t)\Phi c + B(t).$$

But $\Phi(t)$ is a fundamental matrix of $\dot{x} = A(t)x$ and hence satisfies the relation $\dot{\Phi} = A(t)\Phi$, which implies that $\Phi(t)\dot{c} = B(t)$, and conversely.

Furthermore, $W(t) = \det \Phi(t) \neq 0$ for $r_1 < t < r_2$, and therefore $\Phi^{-1}(t)$ exists on (r_1, r_2). The relations (i) and (ii) are then equivalent to

$$\dot{c} = \Phi^{-1}(t)B(t), \qquad c(t_0) = x_0.$$

But the solution of this equation is given by

$$c(t) = x_0 + \int_{t_0}^{t} \Phi^{-1}(s)B(s)\,ds, \qquad r_1 < t < r_2$$

and this gives the desired result.

It should be remarked that the use of the representation above to obtain explicit solutions when $n \geq 3$ is very limited. For it involves finding a fundamental matrix (hence a fundamental system of solutions of $\dot{x} = A(t)x$), then computing an inverse matrix, and finally performing the indicated integration. Finding a fundamental matrix even when $n = 2$ may prove to be a very difficult task.

The value of the representation is that, even knowing only properties of the fundamental matrix $\Phi(t)$ and the behavior of $B(t)$, we may be able to obtain considerable information about the solution $x(t)$. This will become especially clear in Chapter 5, in which stability of solutions of equations of the form above are discussed.

A much simpler representation of solutions of (5), not involving $\Phi^{-1}(t)$, can be given in the case $A(t) = A$, a constant matrix, and we assume $t_0 = 0$. However, we need the following lemma.

LEMMA 2.4.1. *If $\Phi(t)$ is the fundamental matrix of $\dot{x} = Ax$, with A a constant matrix and $\Phi(0) = I$, then $\Phi(t - \alpha) = \Phi(t)\Phi^{-1}(\alpha)$ for every α.*

Proof. Given the real number α, let $\Omega_1(t) = \Phi(t)\Phi^{-1}(\alpha)$; since $\Phi(t)$ satisfies the equation $\dot{\Phi} = A\Phi$, it follows that $\Omega_1(t)$ is the solution of $\dot{\Omega} = A\Omega$ with initial condition $\Omega(\alpha) = I$. But $\Omega_2(t) = \Phi(t - \alpha)$ satisfies $\Omega_2(\alpha) = \Phi(0) = I$, and $\dot{\Omega}_2 = A\Phi(t - \alpha) = A\Omega_2$. By uniqueness, $\Omega_2(t) = \Omega_1(t)$.

Note that even for a specific value of α the above result will not hold for $A = A(t)$ unless $A(t - \alpha) = A(t)$—that is, that α is a period of $A(t)$. Finally, the above two results lead to the following representation for solution of

$$\dot{x} = Ax + B(t), \tag{6}$$

where $A = (a_{ij})$ is a constant matrix and $B(t)$ is continuous on $r_1 < t < r_2$. We assume that $r_1 < 0 < r_2$.

THEOREM 2.4.2. *The solution $x(t)$ of equation* (6) *satisfying $x(0) = x_0$ for $r_1 < 0 < r_2$ is given by*

$$x(t) = \Phi(t)x_0 + \int_0^t \Phi(t - s)B(s)\, ds, \qquad r_1 < t < r_2,$$

where $\Phi(t)$ *is the fundamental matrix of the equation* $\dot{x} = Ax$ *satisfying* $\Phi(0) = I$.

Example: Consider the two-dimensional system

$$\dot{x} = 2x + y + \cos t,$$

$$\dot{y} = 3x + 4y + t.$$

Hence

$$A(t) = A = \begin{pmatrix} 2 & 1 \\ 3 & 4 \end{pmatrix} \quad \text{and} \quad B(t) = \begin{pmatrix} \cos t \\ t \end{pmatrix},$$

so all solutions exist and are uniquely defined on $-\infty < t < \infty$. The solution $\varphi(t) = (x(t),\ y(t))$ satisfying $\varphi(0) = (x_0, y_0)$ is given by

$$\begin{pmatrix} x(t) \\ y(t) \end{pmatrix} = \Phi(t)\begin{pmatrix} x_0 \\ y_0 \end{pmatrix} + \int_0^t \Phi(t-s)\begin{pmatrix} \cos s \\ s \end{pmatrix} ds,$$

where

$$\Phi(t) = \begin{pmatrix} \frac{3}{4}e^t + \frac{1}{4}e^{5t} & -\frac{1}{4}e^t + \frac{1}{4}e^{5t} \\ -\frac{3}{4}e^t + \frac{3}{4}e^{5t} & \frac{1}{4}e^t + \frac{3}{4}e^{5t} \end{pmatrix},$$

as given in the last example of the previous section. The reader may verify, for instance, that the solution satisfying $\varphi(0) = (1, 1)$ is given by $\varphi(t) = (x(t), y(t))$, where

$$x(t) = \frac{5}{8}e^t + \frac{1451}{2600}e^{5t} + \frac{5}{13}\sin t - \frac{11}{26}\cos t + \frac{1}{5}t + \frac{6}{25},$$

$$y(t) = -\frac{5}{8}e^t + \frac{4353}{2600}e^{5t} - \frac{9}{26}\sin t + \frac{3}{13}\cos t - \frac{2}{5}t - \frac{7}{25}.$$

Even the simplest problems lead to rather laborious calculations.

2.5 The nth-Order Linear Equation

We will now apply the results obtained in the previous sections to a consideration of the nth-order linear equation

$$y^{(n)} + a_1(t)y^{(n-1)} + \cdots + a_n(t)y = 0. \tag{7}$$

Here $y = y(t)$ is an unknown scalar function, and the $a_i(t)$, $i = 1, \ldots, n$, are continuous on $r_1 < t < r_2$. The equation is called homogeneous since the right side is zero.

Using the substitution given in the first chapter we let $x_1 = y$ and $x_2 = \dot{y}, \ldots, x_n = y^{(n-1)}$. Then (7) is transformed into the first-order n-dimensional linear system

$$\dot{x} = A(t)x, \tag{8}$$

where

$$A(t) = \begin{pmatrix} 0 & 1 & 0 & \cdots & & 0 \\ 0 & 0 & 1 & \cdots & & 0 \\ \vdots & & & & & \vdots \\ 0 & & & \cdots & 0 & 1 \\ -a_n(t) & & & \cdots & & -a_1(t) \end{pmatrix}.$$

Here $x = x(t) = (x_1(t), \ldots, x_n(t)) = (y(t), \ldots, y^{(n-1)}(t))$ is an unknown vector function. Also note that for (8) the initial condition $x(t_0) = x_0$ is equivalent to an initial condition for (7) of the form

$$y(t_0) = y_0, \; \dot{y}(t_0) = \dot{y}_0, \ldots, y^{(n-1)}(t_0) = y_0^{(n-1)}, \tag{9}$$

where $y_0, \dot{y}_0, \ldots, y_0^{(n-1)}$ are given constants. This is the form of initial conditions for the equation (7).

Equations (7) and (8) are equivalent, since to a solution $y = \psi(t)$ of (7) there corresponds the solution

$$x = \varphi(t) = (\psi(t), \dot{\psi}(t), \ldots, \psi^{(n-1)}(t))$$

of (8). Conversely, given a solution $x = \varphi(t) = (\varphi_1(t), \ldots, \varphi_n(t))$ of (8), there corresponds the solution $y = \varphi_1(t)$ of (7), and (8) implies $\dot{y} = \dot{\varphi}_1 = \varphi_2, \ldots, y^{(n-1)} = \dot{\varphi}_{n-1} = \varphi_n$. We may conclude that given t_0 in (r_1, r_2) and constants $y_0, \dot{y}_0, \ldots, y_0^{(n-1)}$, there exists a unique solution $y = y(t)$ of (7), which is defined on $r_1 < t < r_2$ and satisfies the initial conditions (9).

In view of the equivalence between equations (7) and (8), the discussion of fundamental systems of solutions of (7) and the corresponding Wronskian is considerably simplified. This is one of the great advantages of first having discussed system of the form (8).

DEFINITION: A collection

$$y_1(t), \dots, y_n(t), \qquad r_1 < t < r_2,$$

of solutions of (7) is called a *fundamental system of solutions* of (7) if it is linearly independent.

We can now prove immediately

THEOREM 2.5.1. *A fundamental system of solutions of equation* (7) *exists.*

Proof. By Theorem 2.2.1 a fundamental system of solutions of (8) exists; for example, $\varphi_1(t), \dots, \varphi_n(t)$, where $\varphi_j(t) = (\varphi_{1j}(t), \dots, \varphi_{nj}(t))$. Furthermore, given t_0 in (r_1, r_2) we may assume

$$\varphi_j(t_0) = (0, \dots, 0, \overset{j\text{th place}}{1}, 0, \dots, 0) = e_j, \qquad j = 1, \dots, n.$$

By the correspondence of solutions of (7) and (8) we have

$$\varphi_j(t) = (y_j(t), \dot{y}_j(t), \dots, y_j^{(n-1)}(t))$$

for some solution $y = y_j(t)$ of (7). The collection $y_1(t), \dots, y_n(t)$ are distinct nontrivial solutions, since they satisfy distinct initial conditions and $y_j(t) \equiv 0$ for $r_1 < t < r_2$ would imply that $\varphi_j(t) \equiv 0$, which is impossible.

Finally, if there existed constants $c_1, \dots, c_n$ not all zero such that $\sum_{j=1}^n c_j y_j(t) \equiv 0$ for $r_1 < t < r_2$, then

$$\sum_{j=1}^n c_j \dot{y}_j(t) \equiv 0, \dots, \sum_{j=1}^n c_j y_j^{(n-1)}(t) \equiv 0, \qquad r_1 < t < r_2.$$

This implies that

$$\sum_{j=1}^n c_j \varphi_j(t) \equiv 0, \qquad r_1 < t < r_2,$$

which contradicts the fact that $\varphi_1(t), \dots, \varphi_n(t)$ is a fundamental system of (8).

COROLLARY. *Given any solution $y = y(t)$ of* (7) *and a fundamental system of solutions $y_1(t), \dots, y_n(t)$ of* (7), *there exist constants $c_1, \dots, c_n$ such that*

$$y(t) = \sum_{j=1}^n c_j y_j(t), \qquad r_1 < t < r_2.$$

Proof. Given t_0 in (r_1, r_2), suppose that $y(t_0) = y_0$, $\dot{y}(t_0) = \dot{y}_0, \ldots, y^{(n-1)}(t_0) = y_0^{(n-1)}$. Therefore $(c_1, \ldots, c_n) = c$ must be a solution of the system of equations

$$y_0 = \sum_{j=1}^{n} c_j y_j(t_0),$$

$$\dot{y}_0 = \sum_{j=1}^{n} c_j \dot{y}_j(t_0), \ldots, y_0^{(n-1)} = \sum_{j=1}^{n} c_j y_j^{(n-1)}(t_0).$$

The matrix of coefficients of this system is

$$\Phi(t_0) = \begin{pmatrix} y_1(t_0) & \cdots & y_n(t_0) \\ \dot{y}_1(t_0) & \cdots & \dot{y}_n(t_0) \\ \vdots & & \vdots \\ y_1^{(n-1)}(t_0) & \cdots & y_n^{(n-1)}(t_0) \end{pmatrix}.$$

But $\Phi(t)$ for $r_1 < t < r_2$ is a fundamental matrix, since its columns are a linearly independent set of solutions of (8); hence $\Phi^{-1}(t_0)$ exists. If Y_0 is the vector $(y_0, \dot{y}_0, \ldots, y_0^{(n-1)})$, then the solution is given by $c = \Phi^{-1}(t_0) Y_0$.

We now define the Wronskian of a collection of n solutions of (7).

DEFINITION: Given any collection $y_1(t), \ldots, y_n(t)$ of solutions of (7), then

$$W(t) = \det \begin{pmatrix} y_1 & \cdots & y_n \\ \dot{y}_1 & \cdots & \dot{y}_n \\ \vdots & & \vdots \\ y_1^{(n-1)} & \cdots & y_n^{(n-1)} \end{pmatrix}$$

is called the Wronskian of $y_1(t), \ldots, y_n(t)$.

As before, if the $y_1(t), \ldots, y_n(t)$ are a fundamental system of (7), then the matrix corresponding to $W(t)$ is called a *fundamental matrix*. In any case, note that the columns of the matrix corresponding to $W(t)$ are n solutions of the system (8). We may therefore state immediately a result analogous to Theorem 2.3.1, noting that $\operatorname{tr} A(t) = -a_1(t)$.

THEOREM 2.5.2. *The Wronskian $W(t)$ of any collection $y_1(t), \ldots, y_n(t)$ of solutions of equation* (7) *satisfies the relation*

$$W(t) = W(t_0) \exp\left[-\int_{t_0}^{t} a_1(s)\, ds\right], \qquad r_1 < t_0, t < r_2.$$

Finally, we have the result corresponding to Theorem 2.3.2; the proof is virtually the same.

THEOREM 2.5.3. *A necessary and sufficient condition for $y_1(t), \ldots, y_n(t)$ to be a fundamental system of solutions of equation* (7) *is that*

$$W(t) \neq 0, \qquad r_1 < t < r_2.$$

Examples

(*a*) Given the second-order equation $\ddot{y} + a(t)y = 0$, then for any two solutions $y_1(t)$ and $y_2(t)$ we have

$$W(t) = \det\begin{pmatrix} y_1(t) & y_2(t) \\ \dot{y}_1(t) & \dot{y}_2(t) \end{pmatrix} = \text{constant}, \qquad r_1 < t < r_2.$$

The constant is nonzero if and only if y_1 and y_2 are linearly independent.

(*b*) Given the third-order equation

$$y^{(3)} + \frac{3}{t}\,\ddot{y} - \frac{2}{t^2}\,\dot{y} + \frac{2}{t^3}\,y = 0, \qquad t > 0,$$

then $a_1(t) = 3/t$. The Wronskian of any three solutions $y_1(t)$, $y_2(t)$, $y_3(t)$ satisfies

$$W(t) = W(t_0) \exp\left[-\int_{t_0}^{t} \frac{3}{s}\, ds\right] = W(t_0)\left(\frac{t_0}{t}\right)^3,$$

where t, $t_0 > 0$.

A fundamental system of solutions is given by $y_1(t) = t$, $y_2(t) = t \log t$, and $y_3(t) = 1/t^2$. Therefore, if $t_0 = 1$, we have

$$W(t) = \det\begin{pmatrix} t & t\log t & \dfrac{1}{t^2} \\ 1 & 1+\log t & -\dfrac{2}{t^3} \\ 0 & \dfrac{1}{t} & \dfrac{6}{t^4} \end{pmatrix} \quad \text{and} \quad W(1) = 9,$$

so $W(t) = 9/t^3$, which may be verified directly.

(*c*) The fact that linear independence implies a nonvanishing Wronskian is a property of solutions of linear equations. The functions $y_1(t) = t^3$ and $y_2(t) = |t|^3$ are linearly independent on $-\infty < t < \infty$, but

$$W(t) = \det\begin{pmatrix} t^3 & |t|^3 \\ 3t^2 & 3t\,|t| \end{pmatrix} = 0, \qquad -\infty < t < \infty.$$

Evidently $y_1(t)$ and $y_2(t)$ could not both be solutions near $t = 0$ of a second-order linear equation, since they both satisfy $y(0) = \dot{y}(0) = 0$, yet are distinct. This would violate uniqueness (see Problem 7 of this chapter).

Note that no general methods exist for finding fundamental systems or even one solution of the *n*th-order linear equation. However, a method using power series is available for a rather large class of second-order linear equations; it is discussed in Appendix A.

2.6 The Nonhomogeneous *n*th-Order Linear Equation

To conclude this chapter we will discuss the *n*th-order linear nonhomogeneous equation

$$y^{(n)} + a_1(t)y^{(n-1)} + \cdots + a_n(t)y = b(t), \tag{10}$$

and will use the method of variation of parameters to obtain its solution. Here $y = y(t)$ is an unknown scalar function, and $a_i(t)$, $i = 1, \ldots, n$, and $b(t)$ are continuous on $r_1 < t < r_2$.

The substitution given in the previous section transforms (10) into the first-order n-dimensional system

$$\dot{x} = A(t)x + B(t), \tag{11}$$

where $A(t)$ is given in equation (8) and $B(t) = (0, \ldots, 0, b(t))$. We may conclude that solutions of (10) satisfying initial conditions of the form

$$y(t_0) = y_0, \dot{y}(t_0) = \dot{y}_0, \ldots, y^{(n-1)}(t_0) = y_0^{(n-1)}, \tag{12}$$

where $r_1 < t_0 < r_2$, exist and are uniquely defined on $r_1 < t < r_2$.

To employ the method of variation of parameters, let $y_1(t), \ldots, y_n(t)$ be a fundamental system of solutions of the homogeneous equation corresponding to (10). Then assume that a solution $y(t)$ of (10) satisfying initial conditions (12) can be expressed as

$$y(t) = \sum_{j=1}^{n} c_j(t)y_j(t), \qquad r_1 < t < r_2, \tag{13}$$

where $c_1(t), \ldots, c_n(t)$ are to be determined. This leads to

THEOREM 2.6.1. *The solution of equation* (10) *satisfying initial conditions* (12) *is given by*

$$y(t) = \varphi(t) + W(t_0)^{-1} \sum_{j=1}^{n} y_j(t) \int_{t_0}^{t} \frac{b(s)W_j(s)}{\exp\left[-\int_{t_0}^{s} a_1(u)\,du\right]}\,ds,$$

where

(*i*) $\varphi(t)$ *is the solution satisfying initial conditions* (12) *of the corresponding homogeneous equation,*
(*ii*) $y_1(t), \ldots, y_n(t)$ *are a fundamental system of solutions of the homogeneous equation and* $W(t)$ *is their Wronskian, and*
(*iii*) $W_j(t)$ *is the determinant obtained from* $W(t)$ *by replacing the jth column by* $(0, \ldots, 0, 1)$.

Proof. The matrix $\Phi(t)$ with jth column $(y_j(t), \dot{y}_j(t), \ldots, y^{(n-1)}(t))$ is a fundamental matrix of $\dot{x} = A(t)x$; therefore we assume that $c(t) = (c_1(t), \ldots, c_n(t))$ can be chosen so that $Y(t) = \Phi(t)c(t)$ is a solution of (11). This is equivalent to assuming that $c(t)$ can be chosen so that we have

$$y(t) = \sum_{j=1}^{n} c_j(t)y_j(t),\ \dot{y}(t) = \sum_{j=1}^{n} c_j(t)\dot{y}_j(t), \ldots,$$

$$y^{(n-1)}(t) = \sum_{j=1}^{n} c_j(t)y_j^{(n-1)}(t),$$

and $Y(t) = (y(t),\ \dot{y}(t), \ldots, y^{(n-1)}(t))$ satisfies $Y(t_0) = (y_0, \dot{y}_0, \ldots, y_0^{(n-1)})$.

But now we may proceed as in Theorem 2.4.1 to show that this is equivalent to the relation $\Phi(t)\dot{c} = B(t)$. Hence

$$c(t) = c_0 + \int_{t_0}^{t} \Phi^{-1}(s)B(s)\, ds,$$

so that

$$Y(t) = \Phi(t)c_0 + \int_{t_0}^{t} \Phi(t)\Phi^{-1}(s)B(s)\, ds,$$

where $c_0 = (c_{10}, \ldots, c_{n0}) = c(t_0)$.

The expression $\Phi(t)c_0$ is a solution of the homogeneous system corresponding to (11), so we may write

$$\Phi(t)c_0 = (\varphi(t), \dot{\varphi}(t), \ldots, \varphi^{(n-1)}(t))$$
$$= \left(\sum_{j=1}^{n} c_{j0}\, y_j(t), \sum_{j=1}^{n} c_{j0}\, \dot{y}_j(t), \ldots, \sum_{j=1}^{n} c_{j0}\, y_j^{(n-1)}(t)\right).$$

The relation $Y(t_0) = (y_0, \dot{y}_0, \ldots, y_0^{(n-1)})$ implies that $\varphi(t) = \sum_{j=1}^{n} c_{j0}\, y_j(t)$ is a solution of the homogeneous equation corresponding to (10) satisfying the initial conditions (12).

To determine $c_j(t)$ for $j = 1, \ldots, n$ explicitly, we note that $\Phi(t)\dot{c} = B(t)$ gives the system of equations

$$\begin{aligned}
y_1(t)\dot{c}_1 \quad &+ \ldots + y_n(t)\dot{c}_n = 0,\\
&\vdots\\
y_1^{(n-2)}(t)\dot{c}_1 &+ \ldots + y_n^{(n-2)}(t)\dot{c}_n = 0,\\
y_1^{(n-1)}(t)\dot{c}_1 &+ \ldots + y_n^{(n-1)}(t)\dot{c}_n = b(t).
\end{aligned}$$

The determinant of the coefficients of $\dot{c}_1, \ldots, \dot{c}_n$ is $W(t)$, the Wronskian of the fundamental system $y_1(t), \ldots, y_n(t)$. Using Cramer's rule and the form of the Wronskian given in Theorem 2.5.2, we have

$$\dot{c}_j(t) = \frac{b(t)W_j(t)}{W(t_0)\exp\left[-\int_{t_0}^{t} a_1(s)\,ds\right]}, \qquad c_j(t_0) = c_{j0}, j = 1, \ldots, n,$$

and hence

$$c_j(t) = c_{j0} + W(t_0)^{-1}\int_{t_0}^{t} \frac{b(s)W_j(s)}{\exp\left[-\int_{t_0}^{s} a_1(u)\,du\right]}\,ds.$$

Finally, the relation

$$y(t) = \sum_{j=1}^{n} c_j(t)y_j(t)$$

gives the desired result.

Example: For the second-order equation

$$\ddot{y} + a_1(t)\dot{y} + a_2(t)y = b(t)$$

we have $W_1(t) = -y_2(t)$, $W_2(t) = y_1(t)$ for any fundamental pair of solutions $y_1(t)$ and $y_2(t)$ of the homogeneous equation. If $W(t)$ is their Wronskian, then the solution $y(t)$ satisfying $y(t_0) = y_0$, $\dot{y}(t_0) = \dot{y}_0$ is given by

$$y(t) = c_1 y_1(t) + c_2 y_2(t) - y_1(t)\int_{t_0}^{t} \frac{b(s)y_2(s)}{W(s)}\,ds + y_2(t)\int_{t_0}^{t} \frac{b(s)y_1(s)}{W(s)}\,ds,$$

where c_1 and c_2 are chosen so that $\varphi(t) = c_1 y_1(t) + c_2 y_2(t)$ satisfies the initial conditions.

For instance, the functions $y_1(t) = t^{1/2}$ and $y_2(t) = t^{1/2}\log t$ are a fundamental system of solutions of the equation

$$\ddot{y} + \frac{1}{4t^2}\,y = 0, \qquad t > 0,$$

and $W(t) = 1$. Using the expression above and letting $t_0 = 1$, we may verify that the solution $y(t)$ satisfying $y(1) = 1$, $\dot{y}(1) = 3/2$ of

$$\ddot{y} + \frac{1}{4t^2}\,y = t^{3/2}, \qquad t > 0,$$

is given by

$$y(t) = \tfrac{8}{9}t^{1/2} + \tfrac{2}{3}t^{1/2}\log t + \tfrac{1}{9}t^{7/2}.$$

Problems

1. Given the first-order linear equation

$$\dot{x} = a(t)x + b(t),$$

where $x = x(t)$ is an unknown scalar function and $a(t)$, $b(t)$ are continuous on $r_1 < t < r_2$, show that the solution satisfying $x(t_0) = x_0$ is given by

$$x(t) = \exp\left[\int_{t_0}^{t} a(s)\,ds\right]\left\{x_0 + \int_{t_0}^{t} b(s)\exp\left[-\int_{t_0}^{s} a(u)\,du\right]ds\right\}.$$

Use this to find solutions of the equations

(a) $\dot{x} = x + 2e^t$, $x(0) = 1$.

(b) $\dot{x} = -t^{-1}x - t^{-2}$, $x(1) = 1$.

(c) $\dot{x} = -(t+1)t^{-1}x - 3t^2e^{-t}$, $x(1) = 1$.

2. Verify that $\varphi(t)$ is a solution and then apply the results of the previous problem and of Problem 8, Chapter 1, to find a solution $x(t)$ satisfying $x(t_0) = x_0$ of the following Riccati equations.

(a) $\dot{x} = x + 2t^{-3}x^2 - t^2$, $\varphi(t) = t^2$.

(b) $\dot{x} = x^2 + \cos^2 t + \cos t - 1$, $\varphi(t) = \sin t$.

In (*b*), discuss the behavior of the solution as t approaches $\pm\infty$.

3. Given the inhomogeneous first-order system

$$\dot{x} = 3x - y + 1,$$

$$\dot{y} = 4x - y + t,$$

show that

$$\varphi_1(t) = (e^t, 2e^t), \qquad \varphi_2(t) = (te^t, -e^t + 2te^t),$$

is a fundamental system of solutions of the corresponding homogeneous system. Then find the solution $\varphi(t)$ of the given system satisfying $\varphi(0) = (1, 0)$.

4. Given the nth-order linear equation and a solution $y = y_1(t)$, show that the substitution $y = y_1(t) \int^t u(s)\, ds$ results in a linear equation of order $n - 1$ in $u = u(t)$. This is the method of *reduction of order*.

Use this method to find a second linearly independent solution of the following equations.

(a) $\ddot{y} + 4t\dot{y} + (4t^2 + 2)y = 0,\ y_1(t) = e^{-t^2}$.

(b) $\ddot{y} - (2 \sec^2 t)y = 0,\ y_1(t) = \tan t$.

(c) $\ddot{y} - \dfrac{t+1}{t}\dot{y} - \dfrac{2(t-1)}{t} y = 0,\ y_1(t) = e^{2t}$.

5. (a) Given the equation

$$\ddot{y} + \frac{4}{t}\dot{y} + \frac{2}{t^2} y = \frac{\sin t}{t}, \qquad t > 0,$$

show that $y_1(t) = 1/t$ and $y_2(t) = 1/t^2$ are a fundamental system of solutions of the corresponding homogeneous equation. Then find the solution $y(t)$ of the given equation satisfying $y(1) = 1,\ \dot{y}(1) = 0$.

(b) Proceed as in (a) for the equation

$$\ddot{y} - \frac{2}{t}\dot{y} + y = t^2, \qquad t > 0,$$

where $y_1(t) = \sin t - t \cos t$, $y_2(t) = \cos t + t \sin t$, and the solution $y(t)$ must satisfy $y(\pi/2) = 0,\ \dot{y}(\pi/2) = 1$.

6. Show that if $x_1(t)$ is a nontrivial solution of $\ddot{x} + a(t)x = 0$, where $x = x(t)$ is an unknown scalar function and $a(t)$ is continuous on $r_1 < t < r_2$, then

$$x_2(t) = x_1(t) \int_{t_0}^{t} [x_1(s)]^{-2}\, ds$$

is another linearly independent solution.

7. Given scalar functions $x_1(t)$, $x_2(t)$, continuously differentiable and linearly independent on $a < t < b$, show that if their Wronskian is identically zero on (a, b), then there exists t_0 in (a, b) such that

$$x_1(t_0) = x_2(t_0) = \dot{x}_1(t_0) = \dot{x}_2(t_0) = 0.$$

8. Given scalar functions $u_1(t), \ldots, u_n(t)$, continuous on $a \leq t \leq b$, show that they are linearly dependent if and only if $\det A = 0$, where $A = (a_{ij})$ and

$$a_{ij} = \int_a^b u_i(t)u_j(t)\,dt, \qquad i, j = 1, \ldots, n.$$

9. Using the technique described in Appendix A for the case $s \neq 0$, positive integer, find two linearly independent solutions of the following equations near the regular singular point $z_0 = 0$.

(a) $2z^2 \dfrac{d^2\omega}{dz^2} - z\dfrac{d\omega}{dz} + (z^2 + 1)\omega = 0.$

(b) $2z^2 \dfrac{d^2\omega}{dz^2} - z\dfrac{d\omega}{dz} + (1 - z^2)\omega = 0.$

(c) $z^2 \dfrac{d^2\omega}{dz^2} + z\dfrac{d\omega}{dz} + (z^2 - \alpha^2)\omega = 0$, $\alpha \neq 0$, positive integer (Bessel's equation).

10. The following equations are examples of the case $s =$ positive integer, nonlogarithmic case. Find two linearly independent solutions near the regular singular or ordinary point $z_0 = 0$.

(a) $z\dfrac{d^2\omega}{dz^2} + 2\dfrac{d\omega}{dz} + z^2\omega = 0.$

(b) $z\dfrac{d^2\omega}{dz^2} + (z - 1)\dfrac{d\omega}{dz} - \omega = 0.$

(c) $(1 - z^2)\dfrac{d^2\omega}{dz^2} - 2z\dfrac{d\omega}{dz} + \alpha(\alpha + 1)\omega = 0$, α constant (Legendre's equation).

11. The followir g equations are examples of the case $s = 0$, logarithmic case. Find two linearly independent solutions near the regular singular point $z_0 = 0$.

(a) $z \dfrac{d^2\omega}{dz^2} + \dfrac{d\omega}{dz} - z\omega = 0.$

(b) $z \dfrac{d^2\omega}{dz^2} + \dfrac{d\omega}{dz} + z^2\omega = 0.$

(c) $z^2 \dfrac{d^2\omega}{dz^2} + 3z \dfrac{d\omega}{dz} + (1 + z)\omega = 0.$

12. Using the results of Appendix B, show that if Hill's equation with $T = \pi$ has a nontrivial periodic solution with period $n\pi$ for $n > 2$, but no solution with period π or 2π, then all solutions are periodic with period $n\pi$.

13. Consider the equation

$$\ddot{x} + \varepsilon p(t)x = 0,$$

where $p(t)$ is real-valued, continuous, and periodic with period T, and ε is a parameter.

(a) Show that the fundamental pair of solutions $x_1(t, \varepsilon)$, $x_2(t, \varepsilon)$ satisfying the initial conditions

$$x_1(0, \varepsilon) = 1, \qquad \dot{x}_1(0, \varepsilon) = 0, \qquad x_2(0, \varepsilon) = 0, \qquad \dot{x}_2(0, \varepsilon) = 1$$

can be formally expressed by the power series

$$x_1(t, \varepsilon) = 1 + \varepsilon f_1(t) + \varepsilon^2 f_2(t) + \cdots,$$

$$x_2(t, \varepsilon) = t + \varepsilon g_1(t) + \varepsilon^2 g_2(t) + \cdots,$$

where f_n and g_n satisfy the relations $f_n(0) = g_n(0) = \dot{g}_n(0) = \dot{f}_n(0) = 0$ and

$$\ddot{f}_n + p(t)f_{n-1} = 0, \qquad \ddot{g}_n + p(t)g_{n-1} = 0,$$

for all n, where $f_0(t) = 1$, $g_0(t) = t$.

(b) Assuming the expansion in (a) is valid and letting $\varepsilon = 1$ and $T = \pi$, show that the condition $p(t) < 0$ implies that

$$x_1(\pi, 1) + \dot{x}_2(\pi, 1) > 2.$$

Show that this implies that the corresponding Hill's equation possesses unbounded solutions.

14. Given the first-order linear equation of Problem 1, with $a(t)$ and $b(t)$ periodic of period T, show that it has a periodic solution if and only if $\exp[\int_0^T a(s)\, ds] \neq 1$. Find the periodic solution of the equation $\dot{x} = tx + \sin t$.

CHAPTER

3

The Linear Equation with Constant Coefficients

3.1 The *n*th-Order Linear Equation

In this section we will discuss the equation

$$z^{(n)} + a_1 z^{(n-1)} + \cdots + a_n z = 0, \tag{1}$$

where $z = z(t)$ is an unknown scalar function, possibly complex-valued, and $a_1, \ldots, a_n$ are real or complex constants. The importance of this equation is twofold. First of all, we can immediately determine a fundamental system of solutions, and hence any solution. This is extremely difficult for the general linear equation, whereas for (1) it merely involves the algebraic process of finding the roots of an associated polynomial.

Second, and more important, a large number of physical phenomena can be described in terms of an equation of the form (1) or by a convenient "linearization." For example, the motion of a simple pendulum is governed by the equation $\ddot{\theta} + k \sin \theta = 0$, where $\theta = \theta(t)$ is the angle of displacement and k is a constant. We "linearize" the equation by requiring that $|\theta|$ be small, and consider only the equation $\ddot{\theta} + k\theta = 0$, which is of the form (1) above and for which solutions can be explicitly given.

The question arises, for example, whether the solution $\theta(t)$ of the linear equation adequately describes the motion of a pendulum.

If we rewrite the pendulum equation as $\ddot{\theta} + k\theta + \varepsilon(\theta) = 0$, we are asking the question, "If $|\varepsilon(\theta)|$ is small for $|\theta|$ small, can solutions be described locally in terms of the behavior of solutions of the linear equation?" This type of question will be discussed at length in Chapters 4 and 5, and it will be advantageous to know beforehand about linear equations with constant coefficients.

Equation (1) is a special case of the linear equation discussed in Section 2.5. Therefore, given the initial conditions

$$z(t_0) = z_0, \dot{z}(t_0) = \dot{z}_0, \ldots, z^{(n-1)}(t_0) = z_0^{(n-1)},$$

we know that a unique solution of (1) exists which satisfies them and, since $a_1, \ldots, a_n$ are constants, the solution is defined on $-\infty < t < \infty$. Furthermore, the results proved in Section 2.5 apply here, so we may state immediately the following theorems.

THEOREM 3.1.1. *A fundamental system of solutions of equation* (1) *exists, and every solution can be expressed as a linear combination of the elements of a fundamental system.*

THEOREM 3.1.2. *The Wronskian $W(t)$ of any collection $z_1(t), \ldots, z_n(t)$ of solutions of equation* (1) *satisfies the relation*

$$W(t) = W(0)e^{-a_1 t}, \quad -\infty < t < \infty,$$

and for such a collection to be a fundamental system of solutions a necessary and sufficient condition is that $W(t) \neq 0$, $-\infty < t < \infty$.

To determine explicitly the form of a fundamental system of solutions we will require a few preliminary lemmas and remarks. Recall, first of all, that for the complex number $z = u + iv$ the exponential function e^z is defined by

$$e^z = e^{u+iv} = e^u(\cos v + i \sin v),$$

and the usual laws of exponents hold. Furthermore, $e^z = 1$ if and only if $z = 0$.

LEMMA 3.1.1. *If λ is any real or complex number, then*

$$\frac{d}{dt}(e^{\lambda t}) = \lambda e^{\lambda t}.$$

Proof. It is certainly true for real λ, so we first consider the purely imaginary case $\lambda = iv$. Then

$$e^{\lambda t} = e^{ivt} = \cos vt + i \sin vt$$

and

$$\frac{d}{dt}(e^{\lambda t}) = -v \sin vt + iv \cos vt$$

$$= iv(\cos vt + i \sin vt) = \lambda e^{\lambda t}.$$

Finally, in the case $\lambda = u + iv$,

$$e^{\lambda t} = e^{ut + ivt} = e^{ut}e^{ivt},$$

and the conclusion follows from the product rule for differentiation.

To simplify the discussion we will use operator notation as follows: given any complex or real-valued scalar function $z(t)$, we define the differentiation operator D recursively by

$$Dz = \dot{z}, \qquad D^n z = D(D^{n-1}z) = z^{(n)},$$

where n is any positive integer. By the linear properties of differentiation we may express equation (1) as

$$D^n z + a_1 D^{n-1} z + \cdots + a_n z = L(D)z = 0, \tag{2}$$

where $L(D)$ is the formal polynomial operator given by

$$L(D) = D^n + a_1 D^{n-1} + \cdots + a_n. \tag{3}$$

Again by the linear properties of differentiation it is evident that given any such polynomial operator $L(D)$ and sufficiently differentiable functions $z_1(t)$ and $z_2(t)$, we have for any constants a and b

$$L(D)(az_1 + bz_2) = aL(D)z_1 + bL(D)z_2.$$

Furthermore, given another polynomial operator $M(D)$, then we may define a sum and product by

$$(L + M)(D)z = L(D)z + M(D)z,$$

$$(LM)(D)z = L(D)(M(D)z),$$

where $z = z(t)$ is any sufficiently differentiable function.

Examples

(*a*) If $L(D) = D^2 + 1$ and $M(D) = D^3 + 2D^2 + 1$, then $(L + M)D = D^3 + 3D^2 + 2$ and $(LM)(D) = D^5 + 2D^4 + D^3 + 3D^2 + 1 = (ML)(D)$.

(*b*) The above multiplication is not commutative in the case of nonconstant coefficient polynomial operators. For example, if $L(D) = t^2 D$ and $M(D) = D$, then

$$(LM)(D)z = t^2 D(Dz) = t^2 D^2 z,$$

whereas

$$(ML)(D)z = D(t^2 Dz) = 2tDz + t^2 D^2 z.$$

DEFINITION: Given equation (1), the polynomial

$$L(p) = p^n + a_1 p^{n-1} + \cdots + a_n$$

is called the characteristic polynomial of (1).

The characteristic polynomial of (1) is the key to finding a system of fundamental solutions, as the following theorem and its corollary show.

THEOREM 3.1.3. *If $L(p)$ is any arbitrary polynomial, then $L(D)e^{\lambda t} = L(\lambda)e^{\lambda t}$ for any real or complex number λ.*

Proof. By the definition of the differentiation operator and the result of Lemma 3.1.1, we have for any positive integer k

$$\begin{aligned} D^k(e^{\lambda t}) &= D^{k-1}(De^{\lambda t}) = D^{k-1}(\lambda e^{\lambda t}) \\ &= D^{k-2}(D\lambda e^{\lambda t}) = \cdots = \lambda^k e^{\lambda t}. \end{aligned}$$

If $L(D) = \sum_{j=0}^{n} a_j D^{n-j}$, where we define D^0 by $D^0 z = z$, then

$$L(D)e^{\lambda t} = \sum_{j=0}^{n} a_j D^{n-j}(e^{\lambda t}) = \sum_{j=0}^{n} a_j \lambda^{n-j} e^{\lambda t} = L(\lambda)e^{\lambda t}.$$

COROLLARY. *The function $z(t) = e^{\lambda t}$ is a solution of equation* (1) *if and only if λ is a root of $L(p)$, the characteristic polynomial of* (1).

Proof.

$$L(D)z(t) = L(D)e^{\lambda t} = L(\lambda)e^{\lambda t}, \qquad -\infty < t < \infty,$$

and the right side is identically zero if and only if $L(\lambda) = 0$; that is, λ is a root of $L(p)$.

Example: For the equation

$$z^{(3)} - 4\ddot{z} + 6\dot{z} - 4z = 0,$$

the characteristic polynomial is

$$L(p) = p^3 - 4p^2 + 6p - 4,$$

whose roots are $p = 2,\ 1 \pm i$. Therefore the functions

$$e^{2t},\ e^{(1\pm i)t} = e^t(\cos t \pm i \sin t)$$

are solutions.

Since $L(p)$ is a polynomial of degree n, it has n roots $\lambda_1, \ldots, \lambda_n$, so we can obtain n solutions of (1)—namely, $z_j(t) = e^{\lambda_j t}, j = 1, \ldots, n$. Using these solutions we will obtain a fundamental system of solutions of (1). We will first consider the case in which the roots $\lambda_j, j = 1, \ldots, n$, are distinct.

THEOREM 3.1.4. *If the characteristic polynomial $L(p)$ of equation* (1) *has distinct roots $\lambda_j, j = 1, \ldots, n$, then the functions $z_j(t) = e^{\lambda_j t}, j = 1, \ldots, n$, are a fundamental system of solutions of* (1).

Proof. The functions $z_j(t)$ for $j = 1, \ldots, n$ are solutions by the previous corollary, so we need only prove that they are linearly independent. The simplest way to prove this is to construct their Wronskian $W(t)$ and note that

$$W(0) = \det\begin{pmatrix} 1 & \cdots & 1 \\ \lambda_1 & \cdots & \lambda_n \\ \vdots & & \vdots \\ \lambda_1^{n-1} & \cdots & \lambda_n^{n-1} \end{pmatrix}.$$

This is Vandermonde's determinant and is never zero if the $\lambda_1, \ldots, \lambda_n$ are distinct. From Theorem 3.1.2 it follows that $W(t) \neq 0$ for $-\infty < t < \infty$; hence we have a fundamental system of solutions.

An alternate proof is indicated as follows: suppose $\sum_{j=0}^{n} c_j e^{\lambda_j t} = 0$, $-\infty < t < \infty$, $c_1, \ldots, c_n$ not all zero, and we assume $c_1 \neq 0$. If we multiply the sum by $e^{-\lambda_1 t}$ and differentiate, we obtain

$$\sum_{j=2}^{n} (\lambda_j - \lambda_1) c_j e^{(\lambda_j - \lambda_1)t} = 0, \qquad -\infty < t < \infty.$$

If c_k is the first nonzero c_i, multiply by $e^{-(\lambda_k - \lambda_1)t}$, differentiate, and so on. After a finite number of steps we obtain the relation $Ac_s e^{(\lambda_s - \lambda_r)t} = 0$, where c_s is the last nonzero c_i, $r < s \leq n$, and the constant A depends only on the differences $\lambda_i - \lambda_j$ for $i \neq j$ and is nonzero. Therefore $c_s = 0$, and we procede as before to show that $c_2, \ldots, c_{s-1} = 0$, and finally that $c_1 e^{\lambda_1 t} = 0$, $-\infty < t < \infty$, implies $c_1 = 0$. We conclude that $z_j(t) = e^{\lambda_j t}$, $j = 1, \ldots, n$, are linearly independent.

In the case where the coefficients a_i of $L(p)$ are real, then complex roots of $L(p)$ occur in conjugate pairs. If $\lambda = u + iv$ is a root, so is $\bar{\lambda} = u - iv$, and the corresponding solutions are

$$z_1(t) = e^{(u+iv)t} = e^{ut}(\cos vt + i \sin vt)$$

and (4)

$$z_2(t) = e^{(u-iv)t} = e^{ut}(\cos vt - i \sin vt).$$

Since a linear combination of solutions of (1) is also a solution, this implies that the real-valued functions

$$e^{ut} \cos vt = \tfrac{1}{2}(z_1 + z_2),\ e^{ut} \sin vt = \frac{1}{2i}(z_1 - z_2) \tag{5}$$

are solutions. The following theorem shows that when the coefficients of $L(p)$ are real and the roots are distinct, a fundamental system of real-valued solutions exists.

THEOREM 3.1.5. *Suppose the characteristic polynomial $L(p)$ of equation* (1) *has real coefficients, distinct real roots λ_j, $j = 1, \ldots, k$, and distinct pairs of conjugate complex roots $\lambda_j = u_j + iv_j$, $\lambda_{j+1} = u_j - iv_j$, $j = k + 1, k + 3, \ldots, n - 1$. Then the real-valued functions $z_j(t) = e^{\lambda_j t}$, $j = 1, \ldots, k$, $z_j(t) = e^{u_j t} \cos v_j t$, $z_{j+1}(t) = e^{u_j t} \sin v_j(t)$, $j = k + 1$, $k + 3, \ldots, n - 1$, are a fundamental system of solutions of* (1).

Proof. By the preceding theorem and the remarks above the functions represent n solutions of (1). Suppose a nontrivial linear combination of these solutions was identically zero on $-\infty < t < \infty$. By (3) and (4), we have

$$e^{v_j t}\cos v_j t = \tfrac{1}{2}e^{\lambda_j t} + \tfrac{1}{2}e^{\lambda_{j+1} t},$$

$$e^{v_j t}\sin v_j t = \frac{1}{2i}\,e^{\lambda_j t} - \frac{1}{2i}\,e^{\lambda_{j+1} t},$$

and hence the linear combination is equivalent to an expression of the form $\sum_{j=1}^{n} c_j e^{\lambda_j t} = 0$, $-\infty < t < \infty$. From the foregoing argument, it follows that $c_1, \ldots, c_n = 0$, and this implies that the coefficients of the original linear combination are all zero. Therefore $z_1(t), \ldots, z_n(t)$ are linearly independent.

Examples

(*a*) For the equation

$$z^{(4)} - 4z^{(3)} + 12\ddot{z} + 4\dot{z} - 13z = 0,$$

the characteristic polynomial is

$$L(p) = p^4 - 4p^3 + 12p^2 + 4p - 13,$$

whose roots are $p = \pm 1,\ 2 \pm 3i$. A fundamental system of solutions is then

$$e^t, \qquad e^{-t}, \qquad e^{2t}\cos 3t, \qquad e^{2t}\sin 3t,$$

so any solution is of the form

$$z(t) = c_1 e^t + c_2 e^{-t} + c_3 e^{2t}\cos 3t + c_4 e^{2t}\sin 3t.$$

(*b*) For the equation

$$\ddot{y} + k^2 y = 0, \qquad k \text{ a real constant},$$

the roots of its characteristic polynomial $p^2 + k^2$ are $p = \pm ik$. Therefore a fundamental system of solutions is

$$\sin kt, \qquad \cos kt,$$

and any solution can be written as

$$y(t) = a \sin kt + b \cos kt, \qquad a, b \text{ constant.}$$

If $A = \sqrt{a^2 + b^2}$ and $\alpha = \tan^{-1}(b/a)$, this can be conveniently written as

$$y(t) = A \sin (kt + \alpha),$$

showing the amplitude A and initial phase angle α of the periodic solution.

We now consider the case where the characteristic polynomial $L(p)$ of (1) has multiple roots. Recall that if $L(p) = (p - \lambda)^r g(p)$ and $g(\lambda) \neq 0$, then λ is said to be a root with multiplicity r of $L(p)$, where r is a positive integer. The following theorem gives a complete description of the fundamental system of solutions of (1).

THEOREM 3.1.6. *Let the distinct roots of the characteristic polynomial $L(p)$ of* (1) *be denoted by λ_j with respective multiplicities μ_j, $j = 1, \ldots, k$. Then the functions $t^r e^{\lambda_j t}$, $r = 0, 1, \ldots, \mu_j - 1$, $j = 1, \ldots, k$, are a fundamental system of solutions of* (1).

Proof. Evidently the $e^{\lambda_j t}$ for $j = 1, \ldots, k$ are solutions by the corollary to Theorem 3.1.1. From the properties of polynomials we have

(*i*) $\sum_{j=1}^{k} \mu_j = n$, and

(*ii*) if $\mu_j > 1$, then

$$L(\lambda_j) = L'(\lambda_j) = \cdots = L^{(\mu_j - 1)}(\lambda_j) = 0.$$

Consider the function $f(s, t) = e^{st}$; it has continuous partial derivatives of all orders and therefore, writing $\partial^m e^{st}/\partial t^m = (e^{st})^{(m)}$, we have

$$\frac{\partial^r}{\partial s^r} \frac{\partial^m}{\partial t^m} f(s,t) = \frac{\partial^r}{\partial s^r} (e^{st})^{(m)}$$

$$= \frac{\partial^m}{\partial t^m} \frac{\partial^r}{\partial s^r} f(s,t) = (t^r e^{st})^{(m)}$$

for any positive integers r and m.

We use this fact to show the functions $t^r e^{\lambda_j t}$ are solutions as follows. Consider the function

$$\varphi(s, t) = (e^{st})^{(n)} + a_1(e^{st})^{(n-1)} + \cdots + a_n e^{st} = L(D)e^{st} = e^{st}L(s).$$

By the previous remarks we have

$$\frac{\partial \varphi}{\partial s} = (te^{st})^{(n)} + a_1(te^{st})^{(n-1)} + \cdots + a_n(te^{st})$$

$$= e^{st}[L'(s) + tL(s)].$$

If $s = \lambda_j$ with $\mu_j > 1$, then the right side is zero, which implies that $te^{\lambda_j t}$ is a solution of (1). After r differentiations we would have the expression

$$\frac{\partial^r \varphi}{\partial s^r} = (t^r e^{st})^{(n)} + a_1(t^r e^{st})^{(n-1)} + \cdots + a_n(t^r e^{st})$$

$$= e^{st}[L^{(r)}(s) + rt\, L^{(r-1)}(s) + \cdots + t^r L(s)],$$

and, if $\mu_j > r$, letting $s = \lambda_j$ implies that $t^r e^{\lambda_j t}$ is a solution of (1).

It remains to prove linear independence; we indicate a proof similar to that given in Theorem 3.1.4. Suppose that some nontrivial linear combination of the given functions were identically zero on $-\infty < t < \infty$. This would be equivalent to the existence of polynomials $p_j(t)$, not all zero and of degree $r_j \leq \mu_j - 1, j = 1, \ldots, k$ such that

$$\sum_{j=1}^{k} p_j(t)e^{\lambda_j t} = 0, \qquad -\infty < t < \infty.$$

Assume that $p_1(t) \not\equiv 0$, then multiply the above sum by $e^{-\lambda_1 t}$, and differentiate r_1 times—this annihilates $p_1(t)$. We are left with an expression of the form

$$\sum_{j=2}^{k} q_j(t)e^{(\lambda_j - \lambda_1)t} = 0, \qquad -\infty < t < \infty,$$

where $q_j(t)$ are polynomials of degree r_j. If we proceed as before we eventually arrive at an expression of the form

$$P(t)e^{(\lambda_s - \lambda_r)t} = 0, \qquad -\infty < t < \infty,$$

where $P(t)$ is a nonzero polynomial and $r < s \leq k$. This is a contradiction, since a polynomial has only discrete zeros, and we conclude that the functions

$$t^r e^{\lambda_j t}, \qquad r = 0, 1, \ldots, \mu_j - 1, j = 1, \ldots, k$$

form a fundamental system of solutions of (1).

Finally, we state a result analogous to that of Theorem 3.1.5 for the case where the coefficients of the characteristic polynomial are real. The proof is a slight modification of the previous results and is left to the reader.

THEOREM 3.1.7. *Suppose that the characteristic polynomial $L(p)$ of* (1) *has real coefficients, distinct real roots λ_j with respective multiplicities μ_j, $j = 1, \ldots, k$, and distinct pairs of conjugate complex roots $\lambda_j = u_j + iv_j$, $\lambda_{j+1} = u_j - iv_j$ with respective multiplicities μ_j, $j = m, \ldots, s$. Then the real-valued functions*

$$t^r e^{\lambda_j t}, \qquad j = 1, \ldots, k,$$

$$t^r e^{u_j} \cos v_j t, \quad t^r e^{u_j} \sin v_j t, \qquad j = m, \ldots, s,$$

$$r = 0, 1, \ldots, \mu_j - 1,$$

are a fundamental system of solutions of (1).

Examples

(*a*) The characteristic polynomial of

$$z^{(6)} - 8z^{(5)} + 25z^{(4)} - 32z^{(3)} - \ddot{z} + 40\dot{z} - 25z = 0$$

is

$$L(p) = (p^2 - 1)(p^2 - 4p + 5)^2,$$

and its roots are $p = \pm 1$, and $p = 2 \pm i$ with multiplicity 2. Therefore a fundamental system of solutions is

$$e^t, \quad e^{-t}, \quad e^{2t} \cos t, \quad e^{2t} \sin t, \quad te^{2t} \cos t, \quad te^{2t} \sin t.$$

(*b*) Consider the equation

$$z^{(5)} + 3z^{(4)} + 3z^{(3)} + \ddot{z} = 0.$$

Its characteristic polynomial is

$$L(p) = p^2(p+1)^3,$$

and hence a fundamental system of solutions is

$$1, t, e^{-t}, te^{-t}, t^2e^{-t},$$

and any solution can be expressed as

$$z(t) = c_1 + c_2 t + (c_3 + c_4 t + c_5 t^2)e^{-t},$$

where $c_1, \ldots, c_5$ are constant.

3.2 The Nonhomogeneous nth-Order Linear Equation

We continue the discussion of the nth-order linear equation

$$L(D)z = z^{(n)} + a_1 z^{(n-1)} + \cdots + a_n z = b(t). \tag{6}$$

If $b(t)$ is continuous on $r_1 < t < r_2$, and given initial conditions

$$z(t_0) = z_0, \dot{z}(t_0) = \dot{z}_0, \ldots, z^{(n-1)}(t_0) = z_0^{(n-1)}, \tag{7}$$

where $r_1 < t_0 < r_2$, then by our previous discussion we know that a unique solution $z(t)$ exists satisfying (7) and defined on $r_1 < t < r_2$.

Applying the method of variation of parameters we obtain the following result, corresponding to Theorem 2.6.1.

THEOREM 3.2.1. *The solution of* (6) *satisfying initial conditions* (7) *is given by*

$$z(t) = \varphi(t) + W(0)^{-1} \sum_{j=1}^{n} z_j(t) \int_{t_0}^{t} e^{a_1 s} b(s) W_j(s)\, ds,$$

where

(*i*) *$\varphi(t)$ is the solution satisfying initial conditions* (7) *of the corresponding homogeneous equation,*
(*ii*) *$z_1(t), \ldots, z_n(t)$ are a fundamental system of solutions of the homogeneous equation and $W(t)$ is their Wronskian, and*
(*iii*) *$W_j(t)$ is the determinant obtained from $W(t)$ by replacing the jth column by* $(0, \ldots, 0, 1)$.

Therefore nothing essentially different is obtained from the case of constant coefficients. The only advantage is that we can explicitly determine a fundamental system of solutions for the homogeneous equation, and therefore determine $W_j(t)$.

Example: Consider the equation

$$z^{(3)} - 3\ddot{z} + 2\dot{z} = \log t, \qquad t > 0.$$

A fundamental system of solutions of the corresponding homogeneous equation is

$$z_1(t) = 1, \qquad z_2(t) = e^t, \qquad z_3(t) = e^{2t},$$

and their Wronskian is

$$W(t) = \det\begin{pmatrix} 1 & e^t & e^{2t} \\ 0 & e^t & 2e^{2t} \\ 0 & e^t & 4e^{2t} \end{pmatrix} = 2e^{3t} = W(0)e^{-a_1 t}.$$

Computation gives

$$W_1(t) = e^{3t}, \qquad W_2(t) = -2e^{2t}, \qquad W_3(t) = e^t,$$

and, if $t_0 = 1$, the solution is given by

$$\begin{aligned} z(t) &= \varphi(t) + \tfrac{1}{2}\left[\int_1^t \log s\, ds - 2e^t \int_1^t e^{-s} \log s\, ds \right. \\ &\qquad \left. + e^{2t} \int_1^t e^{-2s} \log s\, ds\right] \\ &= \varphi(t) + \tfrac{1}{2} - \tfrac{1}{2}t + \tfrac{1}{2}t \log t - e^t \int_1^t e^{-s} \log s\, ds \\ &\qquad + \tfrac{1}{2}e^{2t} \int_1^t e^{-2s} \log s\, ds, \end{aligned}$$

where

$$\varphi(t) = c_1 + c_2 e^t + c_3 e^{2t}, \qquad c_1, c_2, c_3 \text{ constant.}$$

It should be noted that the two integrals in the above expression can only be expressed as infinite series, but for practical purposes (that is, numerical calculations), the solution is determined explicitly.

A special method exists for solving (6) when $b(t)$ is of a special form—namely, when it is a solution of a linear equation with constant coefficients. This implies that there exists a polynomial operator

$$M(D) = D^m + b_1 D^{m-1} + \cdots + b_m, \qquad b_i \text{ constant},$$

such that $M(D)b(t) = 0$. By Theorem 3.1.7 it follows that $b(t)$ must be expressible in the form

$$b(t) = \sum_{j=1}^{k} \varphi_j(t) e^{\lambda_j t},$$

where $\varphi_j(t)$, $j = 1, \ldots, k$ are polynomials, and λ_j, $j = 1, \ldots, k$ are the roots of the characteristic polynomial $M(p)$.

The method for solving (6) in this case is called the *annihilator method* or *method of undetermined coefficients*, which we will briefly describe. We are given the relations

$$L(D)z = b(t), \qquad M(D)b(t) = 0,$$

from which it follows that

$$N(D)z = (ML)(D)z = M(D)L(D)z = M(D)b(t) = 0.$$

The left side is a linear equation with constant coefficients, so we can find a fundamental system of solutions $z_1(t), \ldots, z_r(t)$, and the solution of $N(D)z = 0$ is given by

$$\tilde{z}(t) = \sum_{i=1}^{r} c_i z_i(t), \qquad c_i \text{ arbitrary constants.}$$

The relation

$$L(D)\tilde{z}(T) = b(t) \tag{8}$$

will serve to determine by comparison a number of constants, say $c_i = C_i$ for $i = h, \ldots, r$. The solution of (6) is then given by

$$z(t) = \varphi(t) + \sum_{i=h}^{r} C_i z_i(t),$$

where $\varphi(t)$ is the solution of the homogeneous equation corresponding to (6) satisfying the initial conditions.

The evaluation of the constants C_i using relation (8) is not so difficult if we recall that some of the $z_i(t)$ for $i = 1, \ldots, r$ satisfy the relation $L(D)z_i(t) = 0$. This is simply because the roots of the characteristic polynomial $L(p)$ are a subset of the roots of the characteristic polynomial $N(p) = M(p)L(p)$. The following examples illustrate the method.

Examples

(*a*) Consider the equation

$$\ddot{z} - z = e^{2t}(t^2 + 1).$$

We have

$$L(D) = D^2 - 1,$$

and

$$b(t) = e^{2t}(t^2 + 1)$$

satisfies

$$M(D)b(t) = 0,$$

where $M(D) = (D - 2)^3$. It follows that

$$N(D) = (D - 2)^3(D^2 - 1),$$

and the solution of $N(D)z = 0$ is given by

$$\begin{aligned}\tilde{z}(t) &= (c_1 e^t + c_2 e^{-t}) + (c_3 e^{2t} + c_4 te^{2t} + c_5 t^2 e^{2t}) \\ &= \varphi(t) + \beta(t).\end{aligned}$$

Therefore we must choose c_3, c_4, and c_5 so that

$$L(D)\tilde{z} = L(D)\varphi + L(D)\beta = 0 + \ddot{\beta} - \beta = e^{2t}(t^2 + 1).$$

This leads to the relation

$$e^{2t}[3c_5 t^2 + (8c_5 + 3c_4)t + (2c_5 + 4c_4 + 3c_3)] = e^{2t}(t^2 + 1),$$

which, by comparing coefficients, gives

$$c_3 = \tfrac{35}{27}, \qquad c_4 = -\tfrac{8}{9}, \qquad c_5 = \tfrac{1}{3}.$$

The solution is then given by

$$z(t) = c_1 e^t + c_2 e^{-t} + e^{2t}(\tfrac{35}{27} - \tfrac{8}{9}t + \tfrac{1}{3}t^2),$$

where c_1 and c_2 are constants.

(*b*) In the previous example, if we let

$$b(t) = e^{-t}(t+1),$$

then $M(D) = (D+1)^2$ and $N(D) = (D+1)^3(D-1)$, so the solution of $N(D)z = 0$ is given by

$$\tilde{z}(t) = (c_1 e^t + c_2 e^{-t}) + (c_3 te^{-t} + c_4 t^2 e^{-t})$$
$$= \varphi(t) + \beta(t).$$

The relation

$$\ddot{\beta} - \beta = e^{-t}(t+1)$$

gives the result $c_3 = -3/4$, $c_4 = -1/4$, and the solution is then

$$z(t) = c_1 e^t + c_2 e^{-t} - \tfrac{1}{4}e^{-t}(3t + t^2).$$

3.3 The Behavior of Solutions

The solutions of the linear equation

$$L(D)z = z^{(n)} + a_1 z^{(n-1)} + \cdots + a_n z = 0 \tag{9}$$

where $a_1, \ldots, a_n$ are constants, are defined on $-\infty < t < \infty$. In many problems and applications we are interested in the behavior of the solutions as t approaches infinity. This behavior is related to the nature of the roots of the characteristic polynomial $L(p)$ of (9), and we will discuss this relation in this section. Further discussion will be given in Chapters 4 and 5, in which stability of solutions is discussed.

THEOREM 3.3.1. *If all the roots of the characteristic polynomial* $L(p)$ *of* (9) *have negative real parts, then given any solution* $z(t)$ *of* (9) *there exist positive numbers* a *and* M *such that*

$$|z(t)| \le M e^{-at}, \qquad t \ge 0.$$

Hence, $\lim_{t\to\infty} |z(t)| = 0$.

Proof. If $\lambda_j = u_j + iv_j$, $j = 1, \ldots, m$ are the distinct roots of $L(p)$, then by hypothesis $u_j < 0$, so we can find a number $a > 0$ such that $u_j + a < 0$ for $j = 1, \ldots, m$. A solution corresponding to λ_j is of the form $z_i(t) = t^r e^{\lambda_j t}$, and therefore $|z_i(t)e^{at}|$ approaches zero as t approaches infinity.

This implies that there exists a constant $M_i > 0$ such that $|z_i(t)e^{at}| \leq M_i$ for $t \geq 0$, or $|z_i(t)| \leq M_i e^{-at}$ for $t \geq 0$. Any solution $z(t)$ of (9) can be expressed as $z(t) = \sum_{i=1}^n c_i z_i(t)$, where $z_i(t)$ $i = 1, \ldots, n$ are a fundamental system of solutions, and the c_i are constants. If we let $M_0 = \max_i |c_i|$ and $M = M_0 \sum_{i=1}^n M_i$, then for $t \geq 0$

$$
\begin{aligned}
|z(t)| \leq \sum_{i=1}^n |c_i|\,|z_i(t)| &\leq M_0 \sum_{i=1}^n |z_i(t)| \\
&\leq M_0 \left(\sum_{i=1}^n M_i \right) e^{-at} = M e^{-at},
\end{aligned}
$$

which is the desired result.

COROLLARY. *If all roots of $L(p)$ with multiplicity greater than one have negative real parts, and all roots with multiplicity one have nonpositive real parts, then all solutions of* (9) *are bounded for $t \geq 0$.*

Proof. If $\lambda_j = u_j + iv_j$ for $j = 1, \ldots, r$ are the roots with multiplicity one, then $u_j \leq 0$, and a solution $z_j(t)$ corresponding to λ_j satisfies

$$|z_j(t)| = \{(e^{u_j t} \cos v_j t)^2 + (e^{u_j t} \sin v_j t)^2\}^{1/2} = e^{u_j t} \leq 1$$

for $t \geq 0$. The remaining roots λ_j, $j = r + 1, \ldots, n$ have multiplicity greater than one and negative real parts; hence with the same notation as before we have

$$|z(t)| \leq M_0 \sum_{i=1}^n |z_i(t)| \leq M_0 r + \left(\sum_{i=r+1}^n M_i \right) e^{-at}$$

for $t \geq 0$, which implies that $z(t)$ is bounded.

There is only one drawback to the above results—we must know all the roots of the characteristic polynomial $L(p)$ to determine the behavior of solutions of (9). The following theorem gives implicitly a test for the vanishing of solutions as t approaches infinity based on the coefficients of (9). Its proof is not given.

THEOREM 3.3.2 (*Routh-Hurwitz criteria*). *Given the equation* (9) *with* a_i, $i = 1, \ldots, n$, *real, let* $D_1 = a_1$, *and for* $k = 2, \ldots, n$ *let*

$$D_k = \det\begin{pmatrix} a_1 & a_3 & a_5 & \cdots & a_{2k-1} \\ 1 & a_2 & a_4 & \cdots & a_{2k-2} \\ 0 & a_1 & a_3 & \cdots & a_{2k-3} \\ 0 & 1 & a_2 & \cdots & a_{2k-4} \\ \vdots & \vdots & \vdots & & \vdots \\ 0 & 0 & 0 & \cdots & a_k \end{pmatrix},$$

where $a_j = 0$ *if* $j > n$. *Then the roots of* $L(p)$, *the characteristic polynomial of* (9), *have negative real parts if and only if* $D_k > 0$ *for* $k = 1, \ldots, n$.

The test becomes impractical for large n; for $n = 2$, 3, and 4 the results are as follows.

The roots of $L(p)$ have negative real parts if and only if

$n = 2$: a_1 and a_2 are positive,
$n = 3$: a_1, a_2, and a_3 are positive and $a_1 a_2 - a_3 > 0$,
$n = 4$: a_1, a_2, a_3, and a_4 are positive and $a_1 a_2 a_3 - a_3^2 - a_4 a_1^2 > 0$.

Finally, for the roots of $L(p)$ to have negative real parts the following necessary condition is often useful. We assume that the $a_1, \ldots, a_n$ are real.

THEOREM 3.3.3. *If the roots of* $L(p)$, *the characteristic polynomial of* (9), *have negative real parts, then the* $a_1, \ldots, a_n$ *are positive.*

Proof. The polynomial $L(p)$ can be factored into terms of the type $p + a$ and/or $p^2 + bp + c$, a, b, and c real. Since the roots of $L(p)$ have negative real parts, this implies that $a > 0$, $b > 0$, and $c > 0$, which implies that the coefficients of $L(p)$ are positive.

DEFINITION: The characteristic polynomial $L(p)$ of (9) is said to be stable if all its roots have negative real parts.

Examples

(*a*) $L(p) = p^3 + 3p^2 + 2p + 1$ is stable, since its coefficients are positive and

$$a_1 a_2 - a_3 = 3 \cdot 2 - 1 = 5 > 0.$$

Therefore all solutions $z(t)$ of

$$z^{(3)} + 3\ddot{z} + 2\dot{z} + z = 0$$

satisfy $\lim_{t \to \infty} |z(t)| = 0$.

(*b*) $L(p) = p^4 + 6p^3 + 7p^2 + p + 2$ is not stable, since

$$a_1 a_2 a_3 - a_3^2 - a_4 a_1^2 = 6 \cdot 7 \cdot 1 - 1 - 2 \cdot 6^2 = -31 < 0.$$

(*c*) $L(p) = p^5 + 3p^4 - 2p + 1$ is not stable, since $a_4 = 2 - < 0$.

3.4 The First-Order Linear System

We will complete our discussion of the linear equation with constant coefficients by considering the first-order system

$$\dot{x}_i = \sum_{j=1}^{n} a_{ij} x_j, \qquad i = 1, \ldots, n, \tag{10}$$

or

$$\dot{x} = Ax,$$

where $A = (a_{ij})$ is an $n \times n$ matrix with real or complex constant coefficients, and $x = x(t) = (x_1(t), \ldots, x_n(t))$ is an unknown vector function. From our discussion in Section 2.2 it follows that given any initial condition

$$x(t_0) = x_0 = (x_{10}, \ldots, x_{n0}),$$

a unique solution $x(t)$ exists satisfying the initial condition. Furthermore, a fundamental system of solutions $\varphi_1(t), \ldots, \varphi_n(t)$ of (10) exists. We will attempt to give a description of it.

By transposing (10) it may be written in the convenient form

$$\tilde{L}(D)x = 0,$$

where

$$\tilde{L}(D) = \begin{pmatrix} a_{11} - D & a_{12} & \cdots & a_{1n} \\ a_{21} & a_{22} - D & \cdots & a_{2n} \\ \vdots & & & \vdots \\ a_{n1} & a_{n2} & \cdots & a_{nn} - D \end{pmatrix}.$$

Here D is the differential operator and $\tilde{L}(D)$ is a matrix operator. If I is the identity matrix, then we may write

$$\tilde{L}(p) = \begin{pmatrix} a_{11} & \cdots & a_{1n} \\ \vdots & & \vdots \\ a_{n1} & \cdots & a_{nn} \end{pmatrix} - \begin{pmatrix} p & 0 & . & \cdots & 0 \\ 0 & p & 0 & \cdots & 0 \\ \vdots & \vdots & & & \vdots \\ 0 & 0 & . & \cdots & p \end{pmatrix} = A - pI.$$

DEFINITION: The nth-order polynomial $\det \tilde{L}(p) = \det(A - pI)$ is called the characteristic polynomial of the matrix A.

DEFINITION: A number λ that is a root of multiplicity m of $\det(A - pI)$ is called a characteristic root of the matrix A and m is its multiplicity.

Example: If the matrix

$$A = \begin{pmatrix} 4 & -1 & -1 \\ 1 & 2 & -1 \\ 1 & -1 & 2 \end{pmatrix},$$

then

$$A - pI = \begin{pmatrix} 4 - p & -1 & -1 \\ 1 & 2 - p & -1 \\ 1 & -1 & 2 - p \end{pmatrix}$$

and $\det(A - pI) = -(p^3 - 8p^2 + 21p - 18) = -(p - 2)(p - 3)^2$ is the characteristic polynomial of A. Therefore $\lambda = 2$ is a characteristic root of multiplicity 1 and $\lambda = 3$ is one of multiplicity 2.

To show the relation between the characteristic roots of A and the fundamental solutions of (10) we will need some preliminary results. If A is an $n \times n$ matrix and $M(p)$ is any polynomial—for example,

$$M(p) = p^m + a_1 p^{m-1} + \cdots + a_m,$$

then we can construct the corresponding matrix polynomial

$$M(A) = A^m + a_1 A^{m-1} + \cdots + a_m I,$$

where

$$A^k = \overbrace{A \cdot A \cdot \cdots \cdots A}^{k \text{ times}}.$$

If $x(t) = (x_1(t), \ldots, x_n(t))$, then $M(A)x$ makes sense, and if D is the differential operator, then by $M(D)x$ we mean the vector

$$M(D)x = (M(D)x_1, \ldots, M(D)x_n).$$

LEMMA 3.4.1. *If $x = x(t)$ is a solution of* (10), *then so is $x^{(k)}$, its kth derivative for any k. Furthermore, if $M(p)$ is any polynomial, then $M(D)x = M(A)x$.*

Proof. Since $x = x(t)$ is a solution of (10), then $\dot{x} = Ax$ and therefore

$$\frac{d}{dt}(Ax) = \left(\frac{d}{dt}A\right)x + A\frac{dx}{dt} = 0 + A\frac{dx}{dt} = A(Ax),$$

which implies that $y = Ax = \dot{x}$ is a solution of (10), and, furthermore, that $(d/dt)(\dot{x}) = \ddot{x} = A^2x$. It follows by induction that

$$\frac{d}{dt}x^{(k-1)} = x^{(k)} = A^k x$$

is also a solution for any k. From the relation $x^{(k)} = D^k x = A^k x$ and linearity we have $M(D)x = M(A)x$ for any polynomial $M(p)$ and any solution $x = x(t)$ of (10).

THEOREM 3.4.1. *Suppose that λ is a characteristic root of multiplicity m of A and x_0 is a vector satisfying $(A - \lambda I)^m x_0 = 0$. If $x = x(t)$ is the solution of* (10) *satisfying $x(t_0) = x_0$, then $(A - \lambda I)^m x(t) \equiv 0$, $-\infty < t < \infty$.*

Proof. Since $x = x(t)$ is a solution, by the previous lemma we have

(*i*) $(A - \lambda I)^m x(t) = (D - \lambda)^m x(t)$, and
(*ii*) $y = y(t) = (D - \lambda)^m x(t)$ is a solution of (10), since it is a linear combination of x and its derivatives. But by uniqueness $y(t_0) = (A - \lambda I)^m x_0 = 0$ implies that $y(t) \equiv 0$ for $-\infty < t < \infty$.

COROLLARY. *Under the previous hypotheses the solution $x(t)$ of* (10) *satisfying $x(t_0) = x_0$ must be of the form*

$$x(t) = (p_1(t)e^{\lambda t}, \ldots, p_n(t)e^{\lambda t}), \qquad -\infty < t < \infty,$$

where the $p_i(t)$ for $i = 1, \ldots, n$ are polynomials of degree $\leq m - 1$.

Proof. The relation $(D - \lambda)^m x(t) = 0$ for $-\infty\, t < \infty$ is equivalent to the n equations $(D - \lambda)^m x_i = 0$, $i = 1, \ldots, n$, and by the results of Section 3.1 any solution must be of the form indicated.

To complete the description of the fundamental system of solutions of (10) we need the following result from linear algebra.

If $\lambda_1, \ldots, \lambda_k$ are the distinct characteristic roots of A with respective multiplicities $m_1, \ldots, m_k$ so that $\sum_{i=1}^{k} m_i = n$, then to every λ_i there corresponds vectors x_{ij}, $j = 1, \ldots, m_i$, such that

(*i*) *$(A - \lambda_i I)^r x_{ij} = 0$, where $r \leq m_i$, and*
(*ii*) *the collection of vectors x_{ij}, $i = 1, \ldots, k$, $j = 1, \ldots, m_k$ is linearly independent.*

The proof of this result is beyond the algebraic prerequisite intended for this book.

Note that (*i*) implies that $(A - \lambda_i I)^{m_i} x_{ij} = 0$, thus letting $\lambda = \lambda_i$, $m = m_i$, and $x_0 = x(t_0) = x_{ij}$; then the above corollary implies that (10) has a solution

$$x = \varphi(t) = (p_1(t)e^{\lambda_i t}, \ldots, p_n(t)e^{\lambda_i t}), \qquad \varphi(t_0) = x_{ij}, \tag{11}$$

where the $p_s(t)$, $s = 1, \ldots, n$, are polynomials of degree $\leq m_i$. Proceeding in this manner for every x_{ij}, we would obtain a set of n solutions of (10), and these would be linearly independent since their values at $t = t_0$ are the linearly independent set of vectors x_{ij}.

We thus obtain a fundamental system of solutions of (10) of the form (11).

There is only one drawback: given (10), it is generally no easy task to find the vectors x_{ij} described above. The following constructive method, however, obviates this difficulty.

(*a*) Find the characteristic roots of A and their respective multiplicities by computing the characteristic polynomial $\det(A - pI)$.

(*b*) If λ is a characteristic root of A with multiplicity m, first assume a solution of (10) of the form

$$\varphi(t) = (a_1 e^{\lambda t}, \ldots, a_n e^{\lambda t}).$$

Substitute this in (10), which will lead to a system of n equations in the n unknowns $a_1, \ldots, a_n$.

Solve this system to determine any linearly independent solutions, and modify $\varphi(t)$ accordingly.

(*c*) If $m > 1$, next assume a solution of (10) of the form

$$\varphi(t) = ((a_1 + b_1 t)e^{\lambda t}, \ldots, (a_n + b_n t)e^{\lambda t}),$$

and substitute in (10) to obtain $2n$ equations in the $2n$ unknowns $a_1, \ldots, a_n, b_1, \ldots, b_n$. Solve this system to determine any linearly independent solutions and modify $\varphi(t)$ accordingly. Proceed in this manner, finally considering a solution of (10) of the form

$$\varphi(t) = (p_1(t)e^{\lambda t}, \ldots, p_n(t)e^{\lambda t}),$$

where $p_i(t)$ are polynomials of degree $m - 1$ with undetermined coefficients. Proceed as before.

(*d*) Steps (*b*) and (*c*) will result in m linearly independent solutions of (10). Follow these steps for all characteristic roots to determine a fundamental system of solutions of (10).

Examples

(*a*) Consider the system

$$\dot{x} = 5x + 3y, \qquad \dot{y} = -3x - y,$$

and hence

$$A = \begin{pmatrix} 5 & 3 \\ -3 & -1 \end{pmatrix},$$

and $\det(A - pI) = p^2 - 4p + 4 = (p - 2)^2$, so $\lambda = 2$ with multiplicity 2 is the only characteristic root. Assuming a solution of the form

$$\varphi_1(t) = (ae^{2t}, be^{2t})$$

and substituting leads to the relation $b = -a$, a arbitrary, so letting $a = 1$, gives the solution

$$\varphi_1(t) = (e^{2t}, -e^{2t}).$$

Assuming a solution of the form

$$\varphi_2(t) = ((a + bt)e^{2t}, (c + dt)e^{2t})$$

and substituting leads to the relations

$$b = 3a + 3c = -d, \qquad a, c \text{ arbitrary.}$$

Letting $a = 0$, $c = 1$ gives the solution

$$\varphi_2(t) = (3te^{2t}, (1 - 3t)e^{2t}).$$

These are a fundamental system of solutions, since their Wronskian is

$$W(t) = \det\begin{pmatrix} e^{2t} & 3te^{2t} \\ -e^{2t} & (1 - 3t)e^{2t} \end{pmatrix} = e^{4t},$$

so any solution of the system is given by

$$\varphi(t) = (x(t), y(t)) = c_1\varphi_1(t) + c_2\,\varphi_2(t),$$

c_1, c_2 constant.

(*b*) Consider the system

$$\dot{x} = x - 3y, \qquad \dot{y} = 3x + y,$$

and hence

$$A = \begin{pmatrix} 1 & -3 \\ 3 & 1 \end{pmatrix},$$

and $\det(A - pI) = p^2 - 2p + 10$, whose roots are $\lambda = 1 \pm 3i$. In this case we consider a solution of the form

$$\varphi(t) = (ae^t \sin 3t + be^t \cos 3t, ce^t \sin 3t + de^t \cos 3t)$$

and substituting in (10) gives the relation

$$c = b, d = -a, \qquad a, b \text{ arbitrary}.$$

Letting $a = 0$, $b = 1$ we have

$$\varphi_1(t) = (e^t \cos 3t, e^t \sin 3t),$$

and letting $a = 1$, $b = 0$ we obtain

$$\varphi_2(t) = (e^t \sin 3t, -e^t \cos 3t).$$

These are a fundamental system of solutions.

The method of variation of parameters is used to obtain the solution of the nonhomogeneous system

$$\dot{x} = Ax + B(t), A = (a_{ij}), B(t) = (b_1(t), \ldots, b_n(t)).$$

The representation depends on being able to obtain a fundamental matrix $\Phi(t)$ of the corresponding homogeneous system, which we are now able to do. The form of the solution is given in Theorem 2.4.2 and will not be discussed here.

Finally, analogous to Theorem 3.3.1, we have the following result describing the behavior of solutions of (10) as t approaches infinity.

THEOREM 3.4.2. *If the characteristic polynomial* $L(p) = \det(A - pI)$ *of* (10) *is stable, then*

$$\lim_{t \to \infty} \|\varphi(t)\| = 0$$

for any solution $\varphi(t)$ *of* (10).

Proof. If $\varphi_i(t) = (\varphi_{1i}(t), \ldots, \varphi_{ni}(t))$, $i = 1, \ldots, n$, is a fundamental system of solutions of (10), then $\varphi_{ji}(t) = p_{ji}(t)e^{\lambda t}$, $i, j = 1, \ldots, n$, where $p_{ji}(t)$ is a polynomial and λ is a root of $L(p)$. Since the real part of λ is negative, it follows from Theorem 3.3.1 that there exist positive constants M and a such that $|\varphi_{ji}(t)| \le M e^{-at}$, $t > 0$, i, $j = 1, \ldots, n$. This implies that

$$\|\varphi_i(t)\| = \sum_{j=1}^{n} |\varphi_{ji}(t)| \le nM e^{-at}, \qquad t \ge 0,$$

and for any solution $\varphi(t) = \sum_{i=1}^{n} c_i\, \varphi_i(t)$ we then have

$$\|\varphi(t)\| \le \sum_{i=1}^{n} |c_i| \|\varphi_i(t)\| \le K\, e^{-at}, \qquad t \ge 0,$$

where $K = nM \sum_{i=1}^{n} |c_i|$, which gives the desired result.

Problems

1. Determine first if the solutions of the following equations approach zero as t approaches infinity or are bounded. Then describe the real-valued solutions.

(a) $z^{(4)} + 5\ddot{z} + 4z = 0.$

(b) $\ddot{z} + 4\dot{z} + 3z = 0.$

(c) $z^{(3)} + 7\ddot{z} + 16\dot{z} + 12z = 0.$

(d) $z^{(4)} + 2z^{(3)} + 10\ddot{z} = 0.$

(e) $z^{(3)} + \ddot{z} + 4\dot{z} + 4z = 0 \qquad z(0) = \dot{z}(0) = 1,\ \ddot{z}(0) = 2.$

(f) $z^{(4)} + \ddot{z} = 0 \qquad z(0) = \dot{z}(0) = 0,\ \ddot{z}(0) = z^{(3)}(0) = 1.$

2. Using the method of variation of parameters, find the solutions of the following equations.

(a) $\ddot{z} + z = \sec t \tan t.$

(b) $\ddot{z} - 3\dot{z} + 2z = \sin e^{-t}.$

(c) $\ddot{z} - z = t^{-1} - 2t^{-3}.$

(d) $\ddot{z} + z = \cot t.$

3. Using the method of undetermined coefficients, find the solutions of the following equations.

(a) $\ddot{z} - 3\dot{z} + 2z = \sin t.$

(b) $z^{(3)} - \dot{z} = t^2 e^t.$

(c) $\ddot{z} + 4\dot{z} + 3z = t \cos t + e^t.$

(d) $\ddot{z} - z = 4 \sinh t.$

4. (a) Find the general solution of the equation

$$\ddot{y} + q^2 y = A \sin \omega t, \qquad q \ge 0,$$

and show that, if $\omega = q$, then solutions are oscillatory (have an infinite number of zeros on $-\infty < t < \infty$) but become unbounded. This phenomenon is known as *resonance*.

(b) Show that the addition of a *damping term* $2k\dot{y}$ for $k > 0$ on the left side of the previous equation assures that all solutions are bounded regardless of the value of ω.

(c) In the case (b) suppose that $\omega \neq q$ and let

$\varphi = \tan^{-1} 2k\omega/(q^2 - \omega^2), \; -\pi/2 \le \varphi \le \pi/2.$

Express the solution in the form

$y(t) = Y(t) + A\,G \sin(\omega t - \varphi),$

where $Y(t)$ is a solution of the homogeneous equation. The angle φ is called the *steady-state phase angle* and G is a measure of the *gain.*

5. An equation of the form

$(*) \; t^n z^{(n)} + a_1 t^{n-1} z^{(n-1)} + \cdots + a_n z = 0, \qquad a_i \text{ constant},$

is called an Euler equation.
(a) Show that the substitution $t = e^u$ reduces the equation to a nth-order linear equation with constant coefficients. Use this fact to describe a fundamental system of solutions of $(*)$.
(b) For the case $n = 2$ find conditions on the constants a_1 and a_2 that guarantee that all solutions approach zero as t approaches infinity or are bounded.
(c) Find the solutions of the following Euler equations.

(*i*) $t^2\ddot{z} - 4t\dot{z} + 6z = 0.$

(*ii*) $t^2\ddot{z} - 3t\dot{z} + 5z = 0.$

(*iii*) $t^3 z^{(3)} + t\dot{z} - z = 0.$

6. (a) Verify that given $f(t)$ continuous on $-k < t < k$, then

$$z(t) = \int_0^t dt_1 \int_0^{t_1} dt_2 \cdots \int_0^{t_{n-1}} f(t_n)\, dt_n$$

is the solution of the equation

$z^{(n)} = f(t),\; z(0) = \dot{z}(0) = \cdots = z^{(n-1)}(0) = 0.$

(b) Use Dirichlet's formula,

$$\int_a^b dx \int_a^x f(x, y)\, dy = \int_a^b dy \int_y^b f(x, y)\, dx,$$

to show that

$$z(t) = \int_0^t f(u) \frac{(t-u)^{n-1}}{(n-1)!}\, du,$$

and verify that the last expression is the solution for the case $n = 2$.

7. Find a fundamental system of solutions of the following first-order systems.

(a) $\dot{x} = -x + 8y,$
$\dot{y} = x + y.$

(b) $\dot{x} = x + y,$
$\dot{y} = -2x + 3y.$

(c) $\dot{x} = 2x + y,$
$\dot{y} = -x + 4y.$

(d) $\dot{x} = x - 2y - z,$
$\dot{y} = -x + y + z,$
$\dot{z} = x \qquad - z.$

(e) $\dot{x} = x - y + z,$
$\dot{y} = x + y - z,$
$\dot{z} = \quad -y + 2z.$

8. Find a fundamental system of solutions of the corresponding homogeneous system, then use the method of variation of parameters to find solutions of the following systems.

(a) $\dot{x} = -3x + 2y + e^{-t},$
$\dot{y} = -2x + y + 1,$

(b) $\dot{x} = x - y + e^{2t},$
$\dot{y} = -4x + y + t.$

CHAPTER

4

Autonomous Systems and Phase Space

4.1 Introduction

The next two chapters will be devoted to a discussion of how solutions of certain differential equations or systems of differential equations behave, and the notion of stability of solutions will be introduced. The emphasis is not upon finding solutions but upon describing them, and this qualitative study is one of the major aspects of the modern theory of ordinary differential equations.

To begin with we will consider two-dimensional systems of the form

$$\dot{x} = P(x, y), \qquad \dot{y} = Q(x, y), \tag{1}$$

where $x = x(t)$ and $y = y(t)$ are unknown scalar functions, and P and Q together with their first partial derivatives are continuous in some domain Γ of the xy-plane. Such systems are called *autonomous* inasmuch as P and Q do not depend on t. If $z = (x, y)$, then (1) is of the form $\dot{z} = f(z) = (P(x, y), Q(x, y))$, and the hypotheses guarantee existence and uniqueness of solutions by Theorem 1.3.1.

Some reasons for discussing systems of the form (1) are

(*i*) a more complete theory exists than for higher-dimensional systems, and (*ii*) the geometry of the plane and of plane curves is available to illuminate the discussion.

Furthermore, in many cases the analysis of the important second-order autonomous equation

$$\ddot{x} + g(x, \dot{x}) = 0, \qquad x = x(t) \text{ a scalar function,}$$

can be considerably extended by transforming it into the system

$$\dot{x} = y, \qquad \dot{y} = -g(x, y),$$

which is of the form (1).

We begin by giving some simple properties of solutions of (1), and introducing some terminology.

LEMMA 4.1.1. *If* $x = x(t)$, $y = y(t)$, $r_1 < t < r_2$, *is a solution of* (1), *then for any real constant c the functions*

$$x_1(t) = x(t + c), \qquad y_1(t) = y(t + c)$$

are also solutions of (1).

Proof. By the chain rule for differentiation it follows that $\dot{x}_1 = \dot{x}(t + c)$, $\dot{y}_1 = \dot{y}(t + c)$. Since $\dot{x} = P(x(t), y(t))$, $\dot{y} = Q(x(t), y(t))$, replacing t by $t + c$ gives

$$\dot{x}_1 = P(x(t + c), y(t + c)) = P(x_1, y_1),$$

$$\dot{y}_1 = Q(x(t + c), y(t + c)) = Q(x_1, y_1),$$

which implies that x_1 and y_1 are solutions. They are evidently defined on $r_1 - c < t < r_2 - c$.

Remark: The above property does not usually hold for non-autonomous systems; for example, a solution of $\dot{x} = x$, $\dot{y} = tx$ is $x(t) = e^t$, $y(t) = te^t - e^t$, and $\dot{y}(t + c) = (t + c)e^{t+c} \neq tx(t + c)$ unless $c = 0$.

As t varies, a solution $x = x(t)$, $y = y(t)$ of (1) describes parametrically a curve lying in Γ. This curve is called a *trajectory* (orbit, characteristic) of (1).

LEMMA 4.1.2. *Through any point passes at most one trajectory.*

Proof. Let C_1: $x = x_1(t)$, $y = y_1(t)$, and C_2: $x = x_2(t)$, $y = y_2(t)$ be distinct trajectories having a common point

$$(x_0, y_0) = (x_1(t_1), y_1(t_1)) = (x_2(t_2), y_2(t_2)).$$

Then $t_1 \neq t_2$, since otherwise the uniqueness of solutions would be contradicted. By the previous lemma,

$$x(t) = x_1(t + t_1 - t_2), \qquad y(t) = y_1(t + t_1 - t_2)$$

is a solution, and $(x(t_2), y(t_2)) = (x_0, y_0)$ implies that $x(t)$ and $y(t)$ must agree respectively with $x_2(t)$ and $y_2(t)$ by uniqueness. This implies that C_1 and C_2 coincide.

Note carefully the distinction between solutions and trajectories of (1): a trajectory is a curve in Γ that is represented parametrically by more than one solution. Thus $x(t)$, $y(t)$ and $x(t + c)$, $y(t + c)$, $c \neq 0$ represent distinct solutions, but they represent the same curve parametrically.

Example: As α varies between 0 and 2π the functions

$$x(t) = \sin(t + \alpha), \qquad y(t) = \cos(t + \alpha), \qquad -\infty < t < \infty,$$

represent an infinite number of distinct solutions of the system $\dot{x} = y$, $\dot{y} = -x$. They represent the same trajectory, the circle C: $x^2 + y^2 = 1$.

Suppose there exists a solution $x(t) = x_0$, $y(t) = y_0$, $-\infty < t < \infty$, of (1), where x_0 and y_0 are constants. Clearly no trajectory can pass through the point (x_0, y_0), since uniqueness would be violated. Furthermore, we have

$$\dot{x} = 0 = P(x_0, y_0), \qquad \dot{y} = 0 = Q(x_0, y_0),$$

since $x(t)$ and $y(t)$ are solutions. Conversely, if there exists a point (x_0, y_0) in Γ for which $P(x_0, y_0) = Q(x_0, y_0) = 0$, then certainly the functions $x(t) = x_0$, $y(t) = y_0$, $-\infty < t < \infty$, are a solution of (1).

DEFINITION: Any point (x_0, y_0) in Γ at which P and Q both vanish is called a *critical point* of (1). Any other point in Γ is called *regular*.

Other names for critical points are singular points, points of equilibrium, and equilibrium states, and they may be thought of as points where the motion described by (1) is in a state of rest. We call attention to the following kinematic picture.

> Consider the field of vectors $V(x, y) = (P(x, y),\ Q(x, y))$ with (x, y) in Γ. Then (1) describes the motion of a particle (x, y) whose velocity $(\dot{x}, \dot{y})$ is given by $V(x, y)$ at every point in Γ. Trajectories are fixed paths along which the particle moves independent of its starting point, and critical points are points of equilibrium.

Viewed in this way we call Γ the *phase space* of the system (1).

> DEFINITION: A critical point (x_0, y_0) of (1) is called an isolated critical point if there exists a neighborhood of (x_0, y_0) containing no other critical points.

We now introduce the notion of stability of a critical point or, equivalently, stability of the solution $x(t) = x_0$, $y(t) = y_0$, $-\infty < t < \infty$, of (1).

> DEFINITION: Let (x_0, y_0) be an isolated critical point of (1). Then (x_0, y_0) is said to be *stable* if given any $\varepsilon > 0$ there exists $\delta > 0$ such that

(*i*) every trajectory of (1) in the δ-neighborhood of (x_0, y_0) for some $t = t_1$ is defined for $t_1 \le t < \infty$, and
(*ii*) if a trajectory satisfies (*i*) it remains in the ε-neighborhood of (x_0, y_0) for $t > t_1$.

If in addition every trajectory C: $x = x(t)$, $y = y(t)$ satisfying (*i*) and (*ii*) also satisfies

(*iii*) $\lim_{t\to\infty} x(t) = x_0$ and $\lim_{t\to\infty} y(t) = y_0$,

then (x_0, y_0) is said to be *asymptotically stable*. Finally, an isolated critical point that is not stable is said to be *unstable*.

The definition of stability roughly states that (x_0, y_0) is stable if once a trajectory enters a small disc containing (x_0, y_0) it remains within a slightly larger disc for all future time. The above definition is sometimes called stability to the right; a similar definition can be given for stability to the left when t approaches $-\infty$.

Example: The point (0, 0) is the only critical point of the systems

$$(a)\ \dot{x} = y, \qquad (b)\ \dot{x} = -x, \qquad (c)\ \dot{x} = x,$$
$$\dot{y} = -x, \qquad \dot{y} = -y, \qquad \dot{y} = y.$$

In (a) the trajectories are a family of circles C: $x^2 + y^2 = r^2$, $0 < r^2 < \infty$ given by the solutions

$$x(t) = r \sin(t + \alpha), \qquad y(t) = r \cos(t + \alpha).$$

Then (i) and (ii) are satisfied with $r^2 < \delta = \varepsilon$ but (iii) is not; therefore (0, 0) is stable.

In (b) and (c) the trajectories are a family of straight lines C: $y = (y_0/x_0)x$ as well as the lines $x = 0$, $y = 0$, given by the solutions

$$x(t) = x_0\, e^{\pm(t-t_0)}, \qquad y(t) = y_0\, e^{\pm(t-t_0)},$$

not both x_0 and y_0 equal to zero. Here the negative sign is used for (b), the positive sign for (c). For (b) we have (i), (ii), and (iii) satisfied; hence (0, 0) is asymptotically stable. For (c) either $x(t)$ or $y(t)$ or both become infinite as t approaches infinity; hence (0, 0) is unstable.

The phase space of the systems would look like these diagrams, in which arrows denote the direction of increasing time.

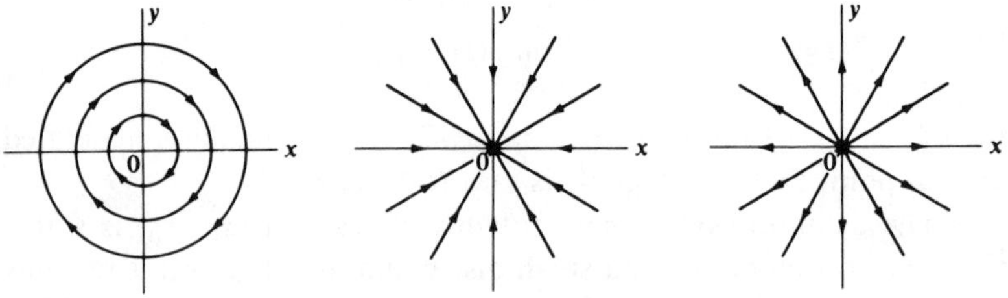

(a) (b) (c)

4.2 Linear Systems—Constant Coefficients

In this section we will consider the linear system

$$\dot{x} = ax + by, \qquad \dot{y} = cx + dy, \tag{2}$$

where a, b, c and d are real constants. Therefore we may let Γ be the entire xy-plane, and so all solutions are uniquely defined on $-\infty < t < \infty$. Hence we can discuss the behavior of trajectories in the *phase plane* of (2).

Why discuss systems of the form (2)? First of all, a complete description of the phase plane can be given, since solutions of (2) can be determined explicitly. Second, many systems can be expressed in the form

$$\dot{x} = ax + by + \varepsilon_1(x, y),$$
$$\dot{y} = cx + dy + \varepsilon_2(x, y).$$

If ε_1 and ε_2 are sufficiently small in the neighborhood of a critical point, we would hope that the behavior of trajectories is locally like that of (2). Thus we need to know about linear systems.

The point $(x_0, y_0) = (0, 0)$ is a critical point of (2), and we will assume there are no other critical points. This is equivalent to assuming that $ad - bc \neq 0$; the case $ad - bc = 0$ is left for the reader to discuss in Problem 2. The characteristic polynomial associated with (2) is

$$\det(A - pI) = \det\begin{pmatrix} a - p & b \\ c & d - p \end{pmatrix}$$
$$= p^2 - (a + d)p + (ad - bc),$$

whose roots are given by

$$\lambda_1, \lambda_2 = \tfrac{1}{2}[(a + d) \pm \sqrt{(a - d)^2 + 4bc}].$$

Then, from our discussion in Section 3.4, solutions are of the form

$$x(t) = f(t)e^{\lambda_i t}, \qquad y(t) = g(t)e^{\lambda_i t},$$

where f and g are polynomials of degree ≤ 1. Since we are only interested in the behavior of trajectories, we will only need to know the nature of the roots λ_i.

To simplify the description of the behavior of trajectories near the critical point (0, 0), it will often be useful to perform a linear transformation of the form

$$\xi = \alpha x + \beta y, \qquad \eta = \gamma x + \delta y, \qquad \alpha\delta - \beta\gamma \neq 0.$$

The point $(x, y) = (0, 0)$ is mapped into $(\xi, \eta) = (0, 0)$, and conversely. Furthermore, such a transformation will only result in a rotation and a magnification or shrinking of trajectories, but will not distort their essential behavior near (0, 0).

Case I: λ_1, λ_2 are real, distinct, and neither is zero:

$(a - d)^2 + 4bc > 0.$

The transformation

$$\xi = cx + (\lambda_1 - a)y, \qquad \eta = cx + (\lambda_2 - a)y$$

transforms (2) into the system

$$\dot{\xi} = \lambda_1 \xi, \qquad \dot{\eta} = \lambda_2 \eta.$$

For instance, since $ad - bc = \lambda_1\lambda_2$ and $\lambda_1 + \lambda_2 = a + d$, we have

$$\begin{aligned}\dot{\xi} &= c\dot{x} + (\lambda_1 - a)\dot{y} = cax + cby + (\lambda_1 - a)(cx + dy)\\ &= \lambda_1 cx + (-\lambda_1\lambda_2)y + \lambda_1 dy = \lambda_1 cx + \lambda_1(\lambda_1 - a)y = \lambda_1 \xi,\end{aligned}$$

and similarly for η.

Therefore, to simplify the discussion we may as well consider the system

$$\dot{x} = \lambda_1 x, \qquad \dot{y} = \lambda_2 y, \qquad \lambda_1 \neq \lambda_2, \qquad \lambda_1\lambda_2 \neq 0, \tag{3}$$

where λ_1 and λ_2 are real. The solutions are of the form

$$x(t) = c_1 e^{\lambda_1 t}, \qquad y(t) = c_2 e^{\lambda_2 t}$$

where c_1 and c_2 are arbitrary real constants.

(*a*) λ_1, λ_2 have the same sign: $ad - bc > 0$;

(*i*) both roots are negative: $a + d < 0$.

If $\lambda_1 < \lambda_2 < 0$, then, as t approaches infinity, (x, y) approaches $(0, 0)$, and y/x, the slope of trajectories near the origin, becomes infinite. If $c_1 = 0$ we have the *rectilinear trajectory*

$$C: x = 0, \; y = c_2 e^{\lambda_2 t},$$

and similarly if $c_2 = 0$.

In this case we say $(0, 0)$ is a *stable node* and the phase plane of (3) looks like the following diagram, in which the arrows denote the direction of increasing time. The diagram will be rotated ninety degrees if $\lambda_2 < \lambda_1 < 0$.

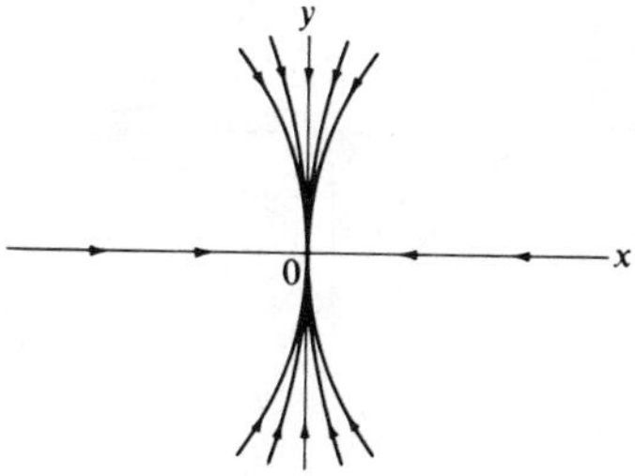

For the corresponding phase plane of (2), the only essential changes in the diagram could consist of a rotation, and possibly the rectilinear trajectories will no longer be perpendicular. Evidently $(0, 0)$ is asymptotically stable.

(*ii*) Both roots are positive: $a + d > 0$.

Then, if $0 < \lambda_1 < \lambda_2$, the diagram is the same with the arrows reversed. In this case $(0, 0)$ is an *unstable node*.

(*b*) λ_1, λ_2 have different sign: $ad - bc < 0$.

If $\lambda_2 < 0 < \lambda_1$, then the rectilinear trajectories are

$$C: x = 0, \; y = c_2 e^{\lambda_2 t},$$

which approaches $(0, 0)$ as t approaches ∞, and

$$C: x = c_1 e^{\lambda_1 t}, \; y = 0,$$

which becomes infinite. If $c_1 > 0$, $c_2 < 0$, then (x, y) approaches $(\infty, 0)$ as t approaches ∞ or (x, y) approaches $(0, -\infty)$ as t approaches $-\infty$. A similar analysis can be made for the other possible values of c_1 and c_2.

In this case we say that (0, 0) is a *saddle point* and is obviously unstable. The phase plane of the system (2) will resemble that below except for possibly a rotation and change of direction of the rectilinear trajectories.

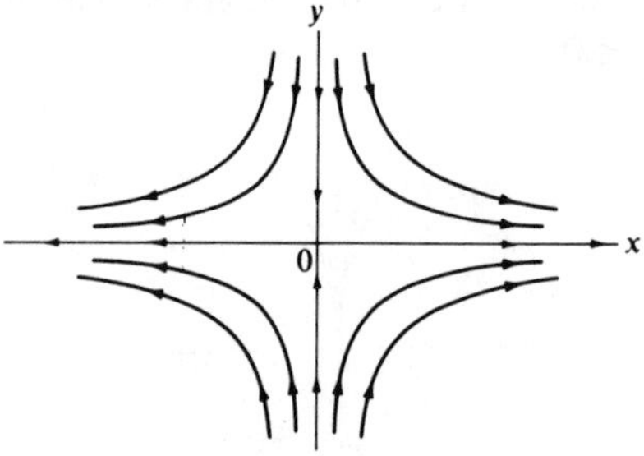

Examples

For the systems

(*a*) $\dot{x} = -3x + y,$

$\dot{y} = 4x - 2y,$

(*b*) $\dot{x} = 2x + y,$

$\dot{y} = x + 2y,$

(*c*) $\dot{x} = 2x + 3y,$

$\dot{y} = x + y,$

$(a - d)^2 + 4bc > 0$ in all cases. The critical point (0, 0) is a stable node for (*a*), since $ad - bc > 0$ and $a + d < 0$; whereas for (*b*) it is an unstable node, since $ad - bc > 0$ and $a + d > 0$. In (*c*) we have $ad - bc < 0$, so (0, 0) is a saddle point.

Case II: λ_1, λ_2 are complex conjugate:

$(a - d)^2 + 4bc < 0.$

We may therefore assume that $\lambda_1 = u + iv$ and $\lambda_2 = u - iv$, where u, v are real numbers. The transformation

$$\xi = cx + (u - a)y, \qquad \eta = vy,$$

transforms (2) into the system

$$\dot{\xi} = u\xi - v\eta, \qquad \dot{\eta} = v\xi + u\eta.$$

Therefore we will consider the system

$$\dot{x} = ux - vy, \qquad \dot{y} = vx + uy, \tag{4}$$

where u and v are real.

(*a*) λ_1, λ_2 are imaginary: $a + d = 0$.

Then $\lambda_1 = iv$, $\lambda_2 = -iv$, $u = 0$ and (4) becomes

$$\dot{x} = -vy, \qquad \dot{y} = vx,$$

whose solutions are

$$x(t) = c_1 \cos(vt + \alpha), \qquad y(t) = c_1 \sin(vt + \alpha),$$

and the trajectories are a family of circles

$$C = x^2 + y^2 = c_1^2.$$

In this case (0, 0) is called a *center* and is stable but not asymptotically stable. The corresponding trajectories for the system (2) will be a family of ellipses. Note in (2) that if $y = 0$, then $\dot{y} = cx$, which indicates that the direction of increasing time is clockwise if $c < 0$ and counterclockwise if $c > 0$.

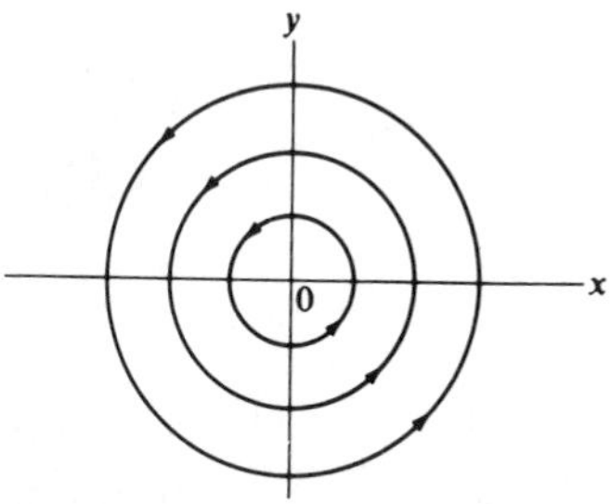

(*b*) λ_1, λ_2 are complex: $a + d \neq 0$.

Then solutions of (4) are

$$x(t) = c_1 e^{ut} \cos(vt + \alpha), \qquad y(t) = c_1 e^{ut} \sin(vt + \alpha),$$

and the trajectories are a family of spirals

$$C: x^2 + y^2 = c_1^2 e^{2ut}.$$

The critical point (0, 0) is called a *spiral point* or *focus* and is asymptotically stable if $a + d = u < 0$, and unstable if $u > 0$. As before, the direction of increasing time is determined by the sign of c. Trajectories have no limiting direction, since $y/z = \tan(vt + \alpha)$ has no limit as t becomes infinite.

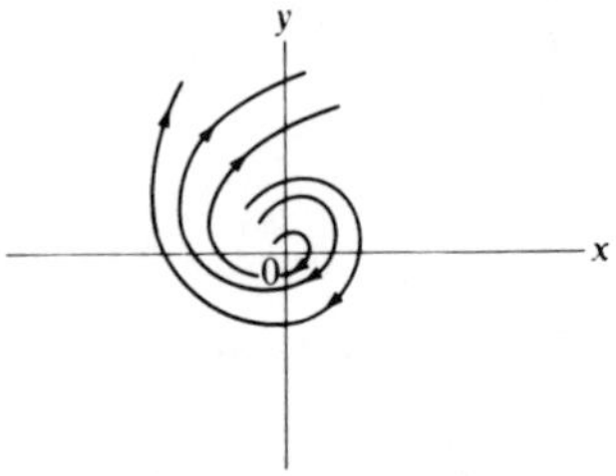

Examples

For the systems

(*a*) $\dot{x} = -x + 3y,$

$\dot{y} = -2x + y,$

(*b*) $\dot{x} = x - 2y,$

$\dot{y} = 3x - 3y,$

(*c*) $\dot{x} = 2x + 2y,$

$\dot{y} = -x + 3y,$

we have $(a - d)^2 + 4bc < 0$. The critical point (0, 0) is a center $(a + d = 0)$, a stable spiral point $(a + d < 0)$, and an unstable spiral point ($a + d > 0$), respectively. For (*a*) and (*c*) the direction of increasing time is clockwise ($c < 0$), whereas for (*b*) it is counterclockwise ($c > 0$).

Case III:

$$\lambda_1 = \lambda_2 = \frac{a+d}{2} \neq 0; (a-d)^2 + 4bc = 0.$$

This is the case of a double root $\lambda_1 = \lambda_2 = u$, where $u = (a+d)/2 \neq 0$ since $ad - bc \neq 0$.

(*a*) A special subcase arises when $b = c = 0$ in (2), which then becomes the system $\dot{x} = ux$, $\dot{y} = uy$, whose solutions are of the form

$$x(t) = c_1 e^{ut}, \qquad y(t) = c_2 e^{ut}.$$

The trajectories are then a family of straight lines C: $y = (c_2/c_1)x$ as well as the lines $x = 0$ and $y = 0$.

Then (0, 0) is called a *proper node* and is asymptotically stable if $a + d < 0$, whereas it is unstable if $a + d > 0$.

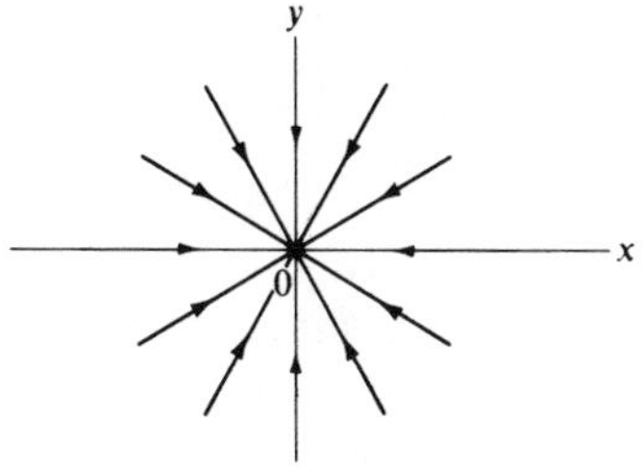

(*b*) In the general case, we may assume that $b \neq 0$. Then the transformation

$$\xi = \frac{a-d}{2b} x + y, \qquad \eta = \frac{1}{b} x,$$

transforms (2) into the system

$$\dot{\xi} = \frac{a+d}{2} \xi, \qquad \dot{\eta} = \xi + \frac{a+d}{2} \eta.$$

(If $b = 0$, $c \neq 0$, then (2) is essentially in this form.) Therefore we may as well consider the system

$$\dot{x} = ux, \qquad \dot{y} = x + uy, \tag{5}$$

whose solutions are

$$x(t) = c_1 e^{ut}, \qquad y(t) = (c_1 t + c_2)e^{ut}.$$

If $c_1 = 0$, the rectilinear trajectory is the y-axis. Otherwise all trajectories are asymptotic to the y-axis, since y/x becomes infinite as t approaches infinity.

In this case the critical point (0, 0) is called a *node* or *improper node*; it is asymptotically stable if $a + d < 0$ and unstable if $a + d > 0$. The phase plane of (5) is sketched below and that of (2) will differ only by a rotation.

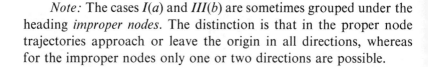

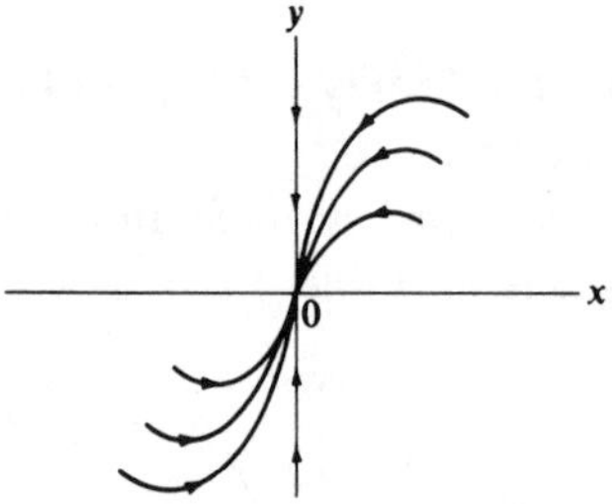

Note: The cases *I*(*a*) and *III*(*b*) are sometimes grouped under the heading *improper nodes.* The distinction is that in the proper node trajectories approach or leave the origin in all directions, whereas for the improper nodes only one or two directions are possible.

Examples

The systems

(*a*) $\dot{x} = -2x$, $\quad$ (*b*) $\dot{x} = 8x - y$,

$\dot{y} = -2y$, $\quad$ $\dot{y} = 4x + 4y$,

represent a stable proper node and an unstable node (improper node), respectively.

From the above analysis we are led to the following result.

THEOREM 4.2.1. *Given the system*

$$\dot{x} = ax + by, \qquad \dot{y} = cx + dy, \qquad ad - bc \neq 0,$$

where a, b, c, and d are real, then (0, 0) *is an isolated critical point and is*

(*i*) *stable if the roots of the characteristic polynomial are purely imaginary,*

(*ii*) *asymptotically stable if the roots have negative real parts, or*

(*iii*) *unstable if the roots have positive real parts.*

4.3 A General Discussion

We return briefly to the general system

$$\dot{x} = P(x, y), \qquad \dot{y} = Q(x, y), \tag{6}$$

where P, Q, and their first partials are continuous in some domain Γ of the xy-plane, and we will assume that Γ is maximal with respect to the last property. The domain Γ can be regarded as the phase space of the system, and hence Γ will contain trajectories of (6) and possibly critical points. To describe it further, we need the following theorem.

THEOREM 4.3.1. *Let* $x = x(t)$, $y = y(t)$ *be a solution of* (6) *defined on a maximum interval of existence* $r_1 < t < r_2$. *If* $x(t_1) = x(t_2)$, $y(t_1) = y(t_2)$, $t_1 \neq t_2$, *then* $r_1 = -\infty$, $r_2 = \infty$, *and the following two cases can occur:*

(*i*) *the solution is an equilibrium state of* (6) *and therefore* $x(t) = x_0$, $y(t) = y_0$, $-\infty < t < \infty$, *where* (x_0, y_0) *is in* Γ, *or*

(*ii*) *the solution is periodic with period* $T > 0$.

Proof. By Lemma 4.1.1 and uniqueness, we have

$$x(t) = x(t + t_1 - t_2), \qquad y(t) = y(t + t_1 - t_2),$$

and since (r_1, r_2) is maximal, this can only occur if $r_1 = -\infty$ and $r_2 = \infty$. Let Π be the set of all periods of the given solutions, and Π is not empty since $t_1 - t_2$ is a period. If c_1 and c_2 are in Π, then $c_1 \pm c_2$ is in Π, and furthermore Π is a closed set. For if c_n, $n = 1, 2, \ldots$, are in Π and $\lim_{n\to\infty} c_n = c$, then by continuity of the solutions $x(t)$ and $y(t)$ it follows that c belongs to Π.

We assert that either Π is the set of all real numbers, and conclusion (*i*) of the theorem follows, or there exists a least positive number T in Π, and conclusion (*ii*) follows. For if there were no least positive number in Π, then given any $\varepsilon > 0$, there exists c in Π such that $0 < c < \varepsilon$. Given any real number r, there exists an integer m such that $|r - mc| \leq c < \varepsilon$, which implies r is in Π by closure, and therefore $\Pi = R$.

Now suppose T is the least positive number in Π. Then we assert that c in Π implies $c = mT$ for some integer m. If not, there would exist some integer n such that $|c - nT| < T$. But since $|c - nT|$ is a period, this contradicts the minimality of T. Therefore Π is the set of all integer multiples of T, and conclusion (*ii*) holds.

From the above theorem and the previous discussion in Section 4.1 we can conclude that the phase space of (6) can only consist of

(*a*) critical points,

(*b*) nonintersecting trajectories, or

(*c*) closed curves called *cycles*, which are trajectories of periodic solutions.

For the linear systems discussed in Section 4.2 the phase space consisted of (*a*) and (*b*), or (*a*) and (*c*) in the case of a center. For a nonlinear system all three may occur, as examples in the following section will show.

If C is any curve in the plane, then by a neighborhood of C we mean a set of points

$$\{(u, v) \mid \|(x, y) - (u, v)\| < \delta, (x, y) \text{ belonging to } C, \delta > 0\}.$$

Using this notion we can introduce the following important class of cycles.

DEFINITION: If K is a cycle of the phase space of (6), then K is called a *limit cycle* if there exists a neighborhood of K such that any trajectory passing through the neighborhood is not a cycle.

Example: The system

$$\dot{x} = y + x(1 - x^2 - y^2), \qquad \dot{y} = -x + y(1 - x^2 - y^2),$$

has the periodic solution

$$x(t) = \cos t, \qquad y(t) = -\sin t,$$

corresponding to the cycle K: $x^2 + y^2 = 1$.

Other solutions may be obtained by using polar coordinates, and therefore

$$\dot{r} = \frac{x\dot{x} + y\dot{y}}{r}, \qquad \dot{\theta} = \frac{x\dot{y} - y\dot{x}}{r^2}.$$

The system then becomes

$$\dot{r} = r(1 - r^2), \qquad \dot{\theta} = -1,$$

and the solutions are

$$r(t) = (1 + ce^{-2t})^{-1/2}, \qquad \theta(t) = -(t - \alpha).$$

These represent a family of spirals that are inside K and tend toward K as t approaches ∞ when $c > 0$. For $c < 0$ they are outside K and tend towards K as t approaches infinity. Therefore the cycle K is a limit cycle.

We may think of a limit cycle K as a closed curve representing an isolated periodic solution of (6), and having the property that trajectories near K spiral toward K, away from K, or both. In this way we can define a stable, unstable or semistable limit cycle.

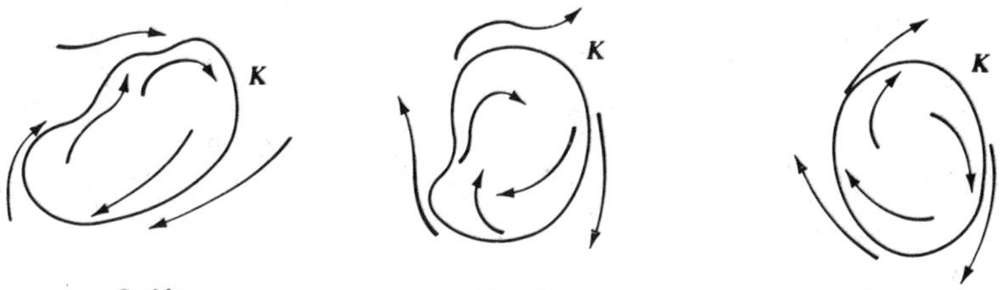

Stable **Unstable** **Semistable**

The investigation of the existence of cycles or limit cycles has led to much fruitful research in this century, but its discussion is beyond the intent of this book.

4.4 Nonlinear Systems

We will now apply the previous analysis given for linear systems in Section 4.2 to systems of the form

$$\begin{aligned}\dot{x} &= P(x, y) = ax + by + \varepsilon_1(x, y),\\ \dot{y} &= Q(x, y) = cx + dy + \varepsilon_2(x, y),\end{aligned} \tag{7}$$

where we assume that

(*i*) P, Q, and their first partials are continuous in some neighborhood of (0, 0),

(*ii*) $ad - bc \neq 0$, and

(*iii*) $\lim_{r\to 0} \dfrac{\varepsilon_i(x, y)}{r} = 0$, $i = 1, 2$, where $r = \sqrt{x^2 + y^2}$.

This implies that (0, 0) is a critical point of (7), and given a system (7) satisfying (*i*), (*ii*), and (*iii*), we will say that (0, 0) is a *simple critical point* of (7).

We will denote by V and $\hat{V}$ the respective vector fields defined by

$$V(x, y) = (P(x, y), Q(x, y)),$$

$$\hat{V}(x, y) = (ax + by, cx + dy),$$

for (x, y) near (0, 0). In view of the assumption (*iii*) we might expect that the phase space of (7) near the origin would resemble that of the "linearized" system

$$\dot{x} = ax + by, \qquad \dot{y} = cx + dy. \tag{8}$$

The following theorem indicates that this is the case. By $\| \ \|$ we will mean the Euclidean norm.

THEOREM 4.4.1. The simple critical point (0, 0) of (7) is isolated and

$$\lim_{r\to 0} \frac{\|V\|}{\|\hat{V}\|} = 1, \qquad \lim_{r\to 0} (\arg V - \arg \hat{V}) = 0.$$

Proof. $\hat{V}$ is continuous and does not vanish on the circle $r = 1$, since (0, 0) is the only critical point of (2). If $v = \inf_{r=1} \| \hat{V} \|$, then $v > 0$ and $\| \hat{V} \| \geq vr$ for all r; hence

$$\lim_{r \to 0} \left\| \frac{V}{\| \hat{V} \|} - \frac{\hat{V}}{\| \hat{V} \|} \right\| \leq \lim_{r \to 0} \frac{\| V - \hat{V} \|}{vr} = 0$$

by assumption (*iii*). The last relation implies that V does not vanish near (0, 0), so the origin is an isolated critical point. The remaining statements follow from the relations

$$\left| \frac{\| V \|}{\| \hat{V} \|} - 1 \right| \leq \left\| \frac{V}{\| \hat{V} \|} - \frac{\hat{V}}{\| \hat{V} \|} \right\|,$$

and

$$\tan^{-1} u - \tan^{-1} v = \tan^{-1} \left[\frac{u - v}{1 + uv} \right].$$

We will now describe the behavior of trajectories of (7) near (0, 0), using the terminology of Section 4.2. To do so we will use polar coordinates. Suppose that $C\colon x = x(t)$, $y = y(t)$ is a trajectory of (7); then we may represent it as

$$C\colon r = r(t), \qquad \omega = \omega(t), \qquad r(t) > 0,$$

where

$$x(t) = r(t) \cos \omega(t), \qquad y(t) = r(t) \sin \omega(t).$$

DEFINITION: Assume there is a neighborhood U of the simple critical point (0, 0) of (7) in which

(*i*) all trajectories are defined on $t_0 < t < \infty$ or $-\infty < t < t_0$ for some t_0;

(*ii*) $\lim_{t \to \infty} r(t) = 0$ or $\lim_{t \to -\infty} r(t) = 0$.

Then (0, 0) is said to be

(*a*) a spiral point if $\lim_{t \to \infty} |\omega(t)| = \infty$ or $\lim_{t \to -\infty} |\omega(t)| = \infty$ for all trajectories in U, (*b*) a node if $\lim_{t\to\infty} \omega(t) = C$ or $\lim_{t\to-\infty} \omega(t) = C$, a constant, for all trajectories in U, or

(*c*) a proper node if it is a node, and for every constant C there is a trajectory satisfying $\lim_{t\to\infty} \omega(t) = C$ or $\lim_{t\to-\infty} \omega(t) = C$.

DEFINITION: The simple critical point (0, 0) of (7) is said to be

(*a*) a center if there exists a neighborhood of (0, 0) containing countably many closed trajectories, each containing (0, 0) and whose diameters tend to zero,

(*b*) a saddle point if there are two trajectories approaching (0, 0) along opposite directions, and all other trajectories close to either of them and to (0, 0) tend away from them as t becomes infinite.

We can now proceed to discuss how the trajectories of (7), with the given assumptions, are related to the trajectories of (8), the linear system, near the simple critical point (0, 0). First of all, if the trajectories of (8) satisfy

$$\lim_{t\to\infty} r(t) = 0 \qquad \text{or} \qquad \lim_{t\to-\infty} r(t) = 0,$$

then so do the trajectories of (7). Hence *asymptotic stability or instability of the origin is preserved.* This result will be proved in Chapter 5, when we discuss a general result for nonautonomous systems.

The following results are also true; proofs are omitted since a detailed proof is required for each individual case.

(*a*) If (0, 0) is a spiral point of (8), it is a spiral point of (7).
(*b*) If (0, 0) is a node of (8), it is a node of (7).
(*c*) If (0, 0) is a saddle point of (8), it is a saddle point of (7).
(*d*) If (0, 0) is a proper node of (8), it is not necessarily a proper node of (7). However, if $\varepsilon_i(x, y)$, $i = 1, 2$, are further restricted—for example,

$$\lim_{r\to 0} \frac{|\varepsilon_i|}{r^{1+\alpha}}$$

is bounded for some $\alpha > 0$—then (0, 0) is a proper node of (7).

(*e*) If (0, 0) is a center of (8), then it is either a center or a spiral point of (7).

In the last case we may think of the elliptical characteristics of the linear system being sufficiently distorted by the $\varepsilon_i(x, y)$ terms to make them spirals. An analysis of these terms or of dr/dt will often resolve the ambiguity.

Examples

(*a*) The motion of a simple pendulum is governed by the equation

$$\ddot{\theta} + 2k\dot{\theta} + q \sin \theta = 0, \qquad k > 0, q > 0,$$

and by the substitution $x = \theta$, $\dot{x} = y$ this becomes the system

$$\dot{x} = y, \qquad \dot{y} = -2ky - q \sin x,$$

which we can rewrite as

$$\dot{x} = y, \qquad \dot{y} = -qx - 2ky + q(x - \sin x).$$

Its critical points are $(\pm n\pi, 0)$, $n = 0, 1, 2, \ldots$, and the term $\varepsilon(x, y) = q(x - \sin x)$ satisfies the required assumptions near $x = 0$, so (0, 0) is a simple critical point.

We therefore consider the system

$$\dot{x} = y, \qquad \dot{y} = -qx - 2ky,$$

which has an isolated singularity at (0, 0). If we assume, for example, that $q > k^2$, then (0, 0) is a stable spiral point of the linear system, and hence is a stable spiral point of the given system.

If we make the change of variable $\theta = \varphi + \pi$, we arrive at the equation

$$\ddot{\varphi} + 2k\dot{\varphi} - q\varphi = 0, \qquad k > 0, q > 0,$$

and a similar analysis shows that (0, 0) is a saddle point of the corresponding system. Therefore $(\pi, 0)$ is a saddle point of the original system, and the phase plane of the pendulum equation conceptually might look like the following diagram.

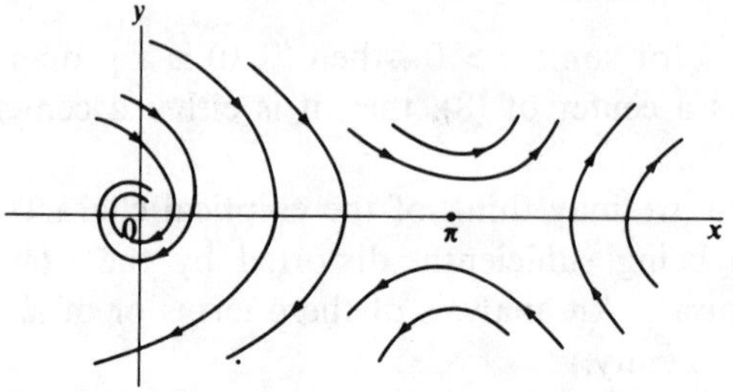

(*b*) The system

$$\dot{x} = y, \qquad \dot{y} = 2x - x^2,$$

has critical points at (0, 0) and (2, 0). The first is a simple critical point ($\varepsilon(x, y) = x^2$) and is a saddle point. By making a change of variable $x = z + 2$, we obtain the system

$$\dot{z} = y, \qquad \dot{y} = -2z - z^2.$$

The point (0, 0) is a center for the corresponding linear system; this is the ambiguous case. By (e) above, a trajectory C passing through the positive x-axis near (0, 0) must intersect the negative x-axis. But the last system is unchanged if we replace y by $-y$ and t by $-t$, which implies that C is closed. Therefore (0, 0) is a center, so (2, 0) is a center for the original system.

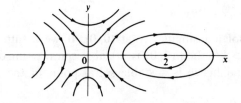

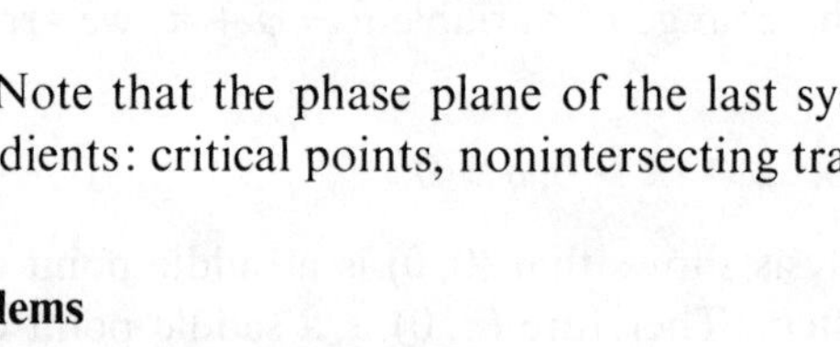

Note that the phase plane of the last system contains all three ingredients: critical points, nonintersecting trajectories, and cycles.

Problems

1. Describe the type and stability of the critical point (0, 0) of the following linear systems.

(a) $\dot{x} = 3x + 4y,$
$\dot{y} = 2x + y.$

(b) $\dot{x} = 3x,$
$\dot{y} = 2x + y.$

(c) $\dot{x} = x + 2y,$
$\dot{y} = -2x + 5y.$

(d) $\dot{x} = 3x - 2y,$
$\dot{y} = 4x - y.$

(e) $\dot{x} = x + 3y,$
$\dot{y} = -6x + 5y.$

(f) $\dot{x} = 3x + y,$
$\dot{y} = -x + y.$

2. Given the linear system

$$\dot{x} = ax + by, \qquad \dot{y} = cx + dy, \qquad a, b, c, d \text{ real},$$

where $ad - bc = 0$, show that the phase plane of the system is one of the following:
(a) a line of critical points with rectilinear trajectories approaching or going away from it,
(b) a line of critical points with rectilinear trajectories parallel to it, or
(c) every point is a critical point.

3. Consider the second-order equation for free oscillations,

$$\ddot{x} + 2k\dot{x} + q^2 x = 0, \qquad k, q \text{ positive constants},$$

and
discuss its solutions with reference to the nature of the critical point (0, 0) of the corresponding linear system,

$$\dot{x} = y, \qquad \dot{y} = -q^2 x - 2ky.$$

4. Find the simple critical points of the following nonlinear systems and describe the local behavior and stability of trajectories. Sketch the phase plane.

(a) $\dot{x} = -4y + 2xy - 8,$
$\dot{y} = 4y^2 - x^2.$

(b) $\dot{x} = x^2 - y^2 - 1,$
$\dot{y} = 2y.$

(c) $\dot{x} = y - x^2 + 2,$
$\dot{y} = 2x^2 - 2xy.$

(d) $\dot{x} = -4x - 2y + 4,$
$\dot{y} = xy.$

(e) $\dot{x} = -x^2 - y^2 + 1,$
$\dot{y} = 2x.$

(f) $\dot{x} = -x^2 - y^2 + 1,$
$\dot{y} = 2xy.$

5. Given the system

$$\dot{x} = -x - y \log^{-1} r, \qquad \dot{y} = -y + x \log^{-1} r,$$

where $r = \sqrt{x^2 + y^2}$, show that (0, 0) is a simple critical point. Then use the polar form to show that it is a spiral point, whereas it is a proper node for the corresponding linear system.

6. Given the system

$$\dot{x} = -y + xr^2 \sin \frac{\pi}{r}, \qquad \dot{y} = x + yr^2 \sin \frac{\pi}{r},$$

show that (0, 0) is the only simple critical point. Use the polar form to show that

(a) the family of circles C_n: $r = 1/n$, $n = 1, 2, \ldots$, are trajectories, and

(b) the above family constitutes the only closed trajectories. Do this by showing that trajectories between any two consecutive circles C_n and C_{n+1} spiral away from or toward the origin, and trajectories outside C_1 become unbounded.

The example shows that, for a center, not all solutions near the origin need be periodic.

7. Show that the origin is a spiral point of the system

$$\dot{x} = -y - xr, \qquad \dot{y} = x - yr, \qquad r = \sqrt{x^2 + y^2},$$

whereas it is a center of the corresponding linear system.

8. Discuss the nature and stability of the simple critical points of the systems corresponding to the following nonlinear equations. Sketch the phase plane of the system.

(a) $\ddot{x} + 6\dot{x} - x^2 + 4x = 0.$

(b) $\ddot{x} + (\dot{x})^3 + x = 0.$

(c) $\ddot{x} + \frac{1}{2}(\dot{x})^2 + 2x^2 - 2 = 0.$

(d) $\ddot{x} + 3|\dot{x}| + 2x = 0.$

(e) $\ddot{x} - \dot{x} + x^2 - 2x = 0.$

(f) $\ddot{x} + \alpha x + \beta x^3 = 0$, α, β constants. Discuss each of the cases $\alpha > 0$ and $\beta < 0$, $\alpha < 0$ and $\beta > 0$, and so on.

CHAPTER

5

Stability for Nonautonomous Equations

5.1 Introduction

We now extend our discussion of stability of solutions to a consideration of the general first-order equation

$$\dot{x} = f(t, x), \tag{1}$$

where $x = x(t) = (x_1(t), \ldots, x_n(t))$ is an unknown n-dimensional vector function, and we assume that

$$f(t, x) = (f_1(t, x), \ldots, f_n(t, x))$$

is defined and continuous in

$$\Gamma = \{(t, x) \mid r_1 < t < \infty, \|x\| < a\}.$$

Recall that if $x = (x_1, \ldots, x_n)$, by the norm $\|x\|$ we mean

$$\|x\| = \sum_{i=1}^{n} |x_i|.$$

A solution (not necessarily unique) of (1) satisfying $x(t_0) = x_0$ will be denoted by

$$x(t) = x(t; t_0, x_0).$$

DEFINITION: Let $x(t) = x(t; t_0, x_0)$ be a solution of (1) satisfying

(*i*) $x(t)$ is defined on $t_0 \leq t < \infty$, and

(*ii*) the point $(t, x(t))$ belongs to Γ for $t \geq t_0$

Then $x(t)$ is said to be *stable* if

(*a*) there exists $\gamma > 0$ such that every solution $x(t; t_0, x_1)$ satisfies (*i*) and (*ii*) whenever $\|x_1 - x_0\| < \gamma$, and

(*b*) given $\varepsilon > 0$ there exists a $\delta > 0, 0 < \delta \leq \gamma$, such that $\|x_0 - x_1\| < \delta$ implies

$$\|x(t; t_0, x_0) - x(t; t_0, x_1)\| < \varepsilon, \qquad t_0 \leq t < \infty.$$

A solution that is not stable is said to be *unstable*.

DEFINITION: The solution $x(t) = x(t; t_0, x_0)$ of (1) is *asymptotically stable* if it is stable and in addition there exists $\rho > 0$, $0 < \rho \leq \gamma$, such that $\|x_0 - x_1\| < \rho$ implies

$$\lim_{t \to \infty} \|x(t; t_0, x_0) - x(t; t_0, x_1)\| = 0.$$

Geometrically, the definitions say that $x(t)$ is stable if any other solution whose initial data is sufficiently close to that of $x(t)$ remains in a "tube" enclosing $x(t)$. If the diameter of the tube approaches zero as t approaches infinity, then $x(t)$ is asymptotically stable. Analogous definitions can be given for t approaching $-\infty$ (stability to the left).

In Chapter 4 we considered the special case when the solution $x(t; t_0, x_0)$ was identically x_0, corresponding to a critical point of the autonomous system.

Examples

(*a*) Every solution of the equation $\dot{x} = 0$ is stable, since

$$\|x(t; t_0, x_0) - x(t; t_0, x_1)\| = \|x_0 - x_1\|, \qquad -\infty < t < \infty,$$

but no solution is asymptotically stable.

(b) Every solution of the equation $\dot{x} = -tx$ is asymptotically stable, since

$$\|x(t; t_0, x_0) - x(t; t_0, x_1)\| = \|x_0 - x_1\| \exp \tfrac{1}{2}(t_0^2 - t^2), \qquad -\infty < t < \infty.$$

(c) The scalar functions

$$x(t) = \tanh (t - t_0 + k), \qquad -\infty < t < \infty,$$

where $k = \tanh^{-1} x_0$, $-1 < x_0 < 1$, are solutions of the equation $\dot{x} = 1 - x^2$, and they are all asymptotically stable since they approach 1 as t approaches ∞. The solution $x = -1$ is not stable, whereas the solution $x = 1$ is asymptotically stable.

(d) The solution $x = 0$ of the equation $\dot{x} = x^2$ is unstable, since for $t_0, x_0 > 0$ the solution

$$x(t; t_0, x_0) = (x_0^{-1} + t_0 - t)^{-1}$$

fails to exist at $t = x_0^{-1} + t_0$.

5.2 Stability for Linear Systems

The problem of stability of solutions of the linear system

$$\dot{x} = A(t)x \tag{2}$$

will first be considered. Here $x = x(t) = (x_1(t), \ldots, x_n(t))$ is an unknown vector function, and the matrix $A(t) = (a_{ij}(t))$ is continuous for $t_0 \le t < \infty$. Recall that the solution of (2) satisfying $x(t_0) = x_0$ is then defined for $t \ge t_0$ and given by $x(t; t_0, x_0) = \Phi(t)x_0$, where $\Phi(t)$ is the fundamental matrix satisfying $\Phi(t_0) = I$.

We will need the notion of the norm of a matrix.

DEFINITION: Given the $n \times n$ matrix $A = (a_{ij})$, then $\|A\|$, the norm of A, is defined by

$$\|A\| = \sum_{i,j=1}^{n} |a_{ij}|.$$

Evidently $\| \ \|$ is a real valued nonnegative function defined on the set of $n \times n$ matrices, and if $A = A(t)$ is continuous, then $\|A(t)\|$ is continuous. In addition it satisfies the properties

(*i*) $\|A + B\| \leq \|A\| + \|B\|$, $\|AB\| \leq \|A\| \, \|B\|$,

(*ii*) $\|cA\| = |c| \, \|A\|$ for any scalar c, and

(*iii*) $\|Ax\| \leq \|A\| \, \|x\|$ for any vector x,

as may be easily verified.

In general, the notions of stability of a solution and boundedness of a solution are independent; for example, the solutions $x = t + x_0$ of $\dot{x} = 1$ are stable but unbounded. However, in the case of linear systems the two notions are equivalent by the following result.

THEOREM 5.2.1. *All solutions of* (2) *are stable if and only if they are bounded.*

Proof. If all solutions of (2) are bounded, then there exists a constant M such that $\|\Phi(t)\| < M$, where $\Phi(t)$ is the fundamental matrix of (2) satisfying $\Phi(t_0) = I$. Given any $\varepsilon > 0$, then $\|x_0 - x_1\| < \varepsilon/M$ implies that

$$\|x(t; t_0, x_0) - x(t; t_0, x_1)\| = \|\Phi(t)(x_0 - x_1)\| \leq M \|x_0 - x_1\| < \varepsilon,$$

and hence all solutions are stable.

Conversely, if all solutions are stable, then the solution $x(t; t_0, 0) \equiv 0$ is stable; therefore, given $\varepsilon > 0$, there exists $\delta > 0$ such that $\|x_1\| < \delta$ implies

$$\|0 - x(t; t_0, x_1)\| = \|\Phi(t)x_1\| < \varepsilon.$$

In particular, we can let x_1 be the vector with $\delta/2$ in the ith place and zero elsewhere. Then

$$\|\Phi(t)x_1\| = \|\Phi_i(t)\| \frac{\delta}{2} < \varepsilon,$$

where $\Phi_i(t)$ is the ith column of $\Phi(t)$, and hence $\|\Phi(t)\| < 2n\varepsilon\delta^{-1} = k$. Therefore for any solution we have

$$\|x(t; t_0, x_0)\| = \|\Phi(t)x_0\| < k \|x_0\|,$$

and hence all solutions are bounded.

We have previously considered in Section 3.4 the linear system $\dot{x} = Ax$, where $A = (a_{ij})$ is a constant matrix. Recall that the characteristic polynomial $\det(A - pI)$ of A is said to be stable if all its roots have negative real parts. We may now rephrase the result of Theorem 3.4.2 as follows.

THEOREM 5.2.2. *If the characteristic polynomial of A is stable, then every solution of $\dot{x} = Ax$ is asymptotically stable.*

Proof. If the characteristic polynomial is stable, then there exist positive constants R and α such that

$$\|\Phi(t)\| \leq Re^{-\alpha t}, \qquad t \geq t_0 \geq 0,$$

where $\Phi(t)$ is the fundamental matrix satisfying $\Phi(t_0) = I$. Since $Re^{-\alpha t}$ is a decreasing function, given $\varepsilon > 0$, then $\|x_0 - x_1\| < \varepsilon R^{-1} e^{\alpha t_0}$ implies

$$\|x(t; t_0, x_0) - x(t; t_0, x_1)\| \leq \|\Phi(t)\| \, \|x_0 - x_1\| \leq Re^{-\alpha t} \|x_0 - x_1\|.$$

The right side is less than ε for $t \geq t_0$ and, furthermore, approaches zero as t approaches ∞, so all solutions are asymptotically stable.

Furthermore, from our discussion of the nature of fundamental systems of solutions of the equation $\dot{x} = Ax$, we can immediately obtain the following result.

THEOREM 5.2.3. *If the multiple roots of the characteristic polynomial of A have negative real parts, and the roots of multiplicity one have nonpositive real parts, then all solutions of $\dot{x} = Ax$ are bounded and hence stable.*

A natural generalization is to consider systems of the form

$$\dot{x} = (A + C(t))x, \tag{3}$$

where $A = (a_{ij})$ is a constant matrix, and the matrix $C(t) = (c_{ij}(t))$ is continuous on $t_0 \leq t < \infty$. We might expect that if the characteristic polynomial of A were stable, then under suitable hypotheses on $C(t)$ the solutions of (3) would be stable. This is the case, and to prove it we will need the following lemma. It is often referred to as Gronwall's inequality and is a most useful tool in the study of stability.

LEMMA 5.2.1. *If the nonnegative scalar functions $u(t)$ and $v(t)$ are continuous on $t_0 \le t < \infty$, α is a nonnegative constant, and*

$$u(t) \le \alpha + \int_{t_0}^{t} v(s)u(s)\,ds$$

for $t \ge t_0$, then

$$u(t) \le \alpha \exp\left[\int_{t_0}^{t} v(s)\,ds\right]$$

for $t \ge t_0$.

Proof. If $\alpha > 0$, then the given inequality implies that

$$\frac{u(t)v(t)}{\alpha + \int_{t_0}^{t} v(s)u(s)\,ds} \le v(t).$$

Integrating both sides from t_0 to t gives

$$\log\left[\alpha + \int_{t_0}^{t} v(s)u(s)\,ds\right] - \log\alpha \le \int_{t_0}^{t} v(s)\,ds,$$

which implies that

$$u(t) \le \alpha + \int_{t_0}^{t} v(s)u(s)\,ds \le \alpha \exp\left[\int_{t_0}^{t} v(s)\,ds\right].$$

If $\alpha = 0$, the result holds for every $\alpha_1 > 0$, and as α_1 approaches zero this implies that $u(t)$ is identically zero and the inequality is trivially satisfied.

Using the lemma we can now prove the following.

THEOREM 5.2.4. *If the characteristic polynomial of A is stable, the matrix $C(t)$ is continuous on $0 \le t < \infty$, and*

$$\int_{0}^{\infty} \|C(t)\|\,dt < \infty,$$

then all solutions of $\dot{x} = (A + C(t))x$ are asymptotically stable.

Proof. Using the expression given in Theorem 2.4.2 with $B(t) = C(t)x(t)$, the solution $x(t) = x(t; 0, x_0)$ of the equation must satisfy the expression

$$x(t) = \Phi(t)x_0 + \int_0^t \Phi(t-s)C(s)x(s)\,ds.$$

Here $\Phi(t)$ is the fundamental matrix of the equation $\dot{x} = Ax$, and $\Phi(0) = I$. Furthermore, the hypotheses imply there exists positive constants R and α such that $\|\Phi(t)\| \leq Re^{-\alpha t}$ for $t \geq 0$, and therefore

$$\|x(t)\|e^{\alpha t} \leq R\|x_0\| + \int_0^t R\|C(s)\|\,\|x(s)\|\,e^{\alpha s}\,ds.$$

We may now apply Lemma 5.2.1, which for $t \geq 0$ gives

$$\|x(t)\|\,e^{\alpha t} \leq R\,\|x_0\|\exp\left[R\int_0^t \|C(s)\|\,ds\right]$$

$$\leq R\,\|x_0\|\exp\left[R\int_0^\infty \|C(s)\|\,ds\right] = M < \infty.$$

This implies that all solutions are bounded and hence are stable, and furthermore that they approach zero as t approaches ∞. Since the difference of any two solutions of a linear system is also a solution, this implies that all solutions are asymptotically stable.

COROLLARY 1. *The conclusion of the theorem holds if the characteristic polynomial of A is stable, $C(t)$ is continuous on $0 \leq t < \infty$, and $\|C(t)\| < c$ for $t \geq 0$ with c sufficiently small.*

Proof. Proceeding as above we arrive at the inequality

$$\|x(t)\|\,e^{\alpha t} \leq R\,\|x_0\|\exp\left[R\int_0^t \|C(s)\|\,ds\right]$$

$$\leq R\,\|x_0\|\,e^{Rct}, \qquad t \geq 0,$$

or

$$\|x(t)\| \leq R\,\|x_0\|e^{(Rc-\alpha)t}, \qquad t \geq 0.$$

If c is small enough, so that $Rc - \alpha < 0$, the result follows.

COROLLARY 2. *If all solutions of* $\dot{x} = Ax$ *are bounded,* $C(t)$ *is continuous on* $0 \le t < \infty$ *and* $\int_0^\infty \|C(t)\| \, dt < \infty$, *then all solutions of* $\dot{x} = (A + C(t))x$ *are bounded and hence stable.*

Proof. If $\Phi(t)$ is the fundamental matrix of the equation $\dot{x} = Ax$ such that $\Phi(0) = I$, then by hypothesis there exists a constant K such that $\|\Phi(t)\| \le K$, $0 \le t < \infty$. It follows that for $x(t) = x(t; 0, x_0)$ we have

$$\|x(t)\| \le K\|x_0\| + \int_0^t K\|C(s)\| \, \|x(s)\| \, ds, \qquad t \ge 0,$$

and therefore

$$\begin{aligned}\|x(t)\| &\le K\|x_0\| \exp\left[K \int_0^t \|C(s)\| \, ds\right] \\ &\le K\|x_0\| \exp\left[K \int_0^\infty \|C(s)\| \, ds\right] = M < \infty.\end{aligned}$$

Therefore all solutions are bounded, and hence stable by Theorem 5.2.1.

Example: The second-order equations

$$(a)\ \ddot{x} + x = 0, \qquad (b)\ \ddot{x} - \frac{2\dot{x}}{t + t_0} + x = 0,\ t_0 > 0,$$

correspond respectively to the systems

$$(a')\ \dot{x} = Ax, \qquad (b')\ \dot{x} = (A + C(t))x,$$

where

$$A = \begin{pmatrix} 0 & 1 \\ -1 & 0 \end{pmatrix} \quad \text{and} \quad C(t) = \begin{pmatrix} 0 & 0 \\ 0 & \dfrac{2}{t + t_0} \end{pmatrix}.$$

A fundamental system of solutions of (a) is $\sin t$, $\cos t$ so all solutions of (a) and (a') are bounded. A fundamental system of solutions of (b) is $\sin t - (t + t_0)\cos t$, $\cos t + (t + t_0)\sin t$, so a nontrivial solution of (b) and (b') is unbounded.

For $t \geq 0$, we have

$$\|C(t)\| = \frac{2}{t + t_0} \leq \frac{2}{t_0},$$

which can be made as small as desired by the choice of t_0. This example confirms that the hypothesis that $\|C(t)\|$ sufficiently small is not enough to insure that solutions of $\dot{x} = (A + C(t))x$ are bounded when those of $\dot{x} = Ax$ are bounded. Note that

$$\int_0^t \|C(s)\| \, ds = \log\left(\frac{t + t_0}{t_0}\right)^2,$$

which becomes infinite as t approaches ∞.

5.3 Two Results for Nonlinear Systems

We will first consider the nonlinear equation

$$\dot{x} = A(t)x + f(t, x), \tag{4}$$

where $A(t) = (a_{ij}(t))$ is a continuous $n \times n$ matrix defined on $0 \leq t < \infty$, and the vector function

$$f(t, x) = (f_1(t, x), \ldots, f_n(t, x))$$

satisfies

(i) $f(t, x)$ is continuous for $\|x\| < a$, $0 \leq t < \infty$, and
(ii) $\lim_{\|x\| \to 0} \|f(t, x)\|/\|x\| = 0$ uniformly with respect to t; that is, $\|f(t, x)\| = o(\|x\|)$ uniformly in t as $\|x\|$ approaches zero.

Condition (i) assures local existence, but not necessarily uniqueness of solutions, and (ii) implies $f(t, 0) = 0$; hence $x(t) \equiv 0$, $0 \leq t < \infty$, is a solution of (4).

In view of the condition (ii), we might argue heuristically that since $A(t)x + f(t, x)$ is very nearly like $A(t)x$ for $\|x\|$ near zero, then if solutions of $\dot{x} = A(t)x$ approach zero as t approaches ∞, so do those of (4). This would be equivalent to asserting that the solution $x(t) \equiv 0$ of (4) is asymptotically stable. Such an argument is correct in the case $A(t) = A$, a real constant matrix, as the following theorem shows.

THEOREM 5.3.1. *If A is an $n \times n$ constant matrix whose characteristic polynomial is stable, and $f(t, x)$ satisfies conditions (i) and (ii) above, then the solution $x(t) \equiv 0$ of the system*

$$\dot{x} = Ax + f(t, x)$$

is asymptotically stable.

Proof. We first show that the solution $x(t) = x(t; 0, x_0)$ is defined on $0 \leq t < \infty$ when x_0 is near zero. If $\Phi(t)$ is the fundamental matrix of the system $\dot{x} = Ax$ such that $\Phi(0) = I$, then by hypothesis there exist positive constants R and α such that $\|\Phi(t)\| \leq Re^{-\alpha t}$ for $t \geq 0$.

Since A is a constant matrix, the solution $x(t)$ must satisfy the relation

$$x(t) = \Phi(t)x_0 + \int_0^t \Phi(t-s)f(s, x(s))\, ds,$$

which implies

$$\|x(t)\|\, e^{\alpha t} \leq R\, \|x_0\| + \int_0^t Re^{\alpha s} \|f(s, x(s))\|\, ds.$$

The first relation and hence the second is certainly valid for t in any interval $[0, T)$ for which $\|x(t)\| < a$ if we assume $\|x_0\| < a$.

From condition (ii) it follows that given any $m > 0$ there exists $d > 0$ such that for $t \geq 0$ and $\|x\| < d$ we have $\|f(t, x)\| \leq m\, \|x\|$. If we assume $\|x_0\| < d$, then by continuity of $x(t)$ there exists $t_1 > 0$ such that $\|x(t)\| < d$ for $0 \leq t < t_1$. Therefore

$$\|x(t)\|\, e^{\alpha t} \leq R\, \|x_0\| + \int_0^t mRe^{\alpha s}\, \|x(s)\|\, ds$$

for $0 \leq t < t_1$. By Lemma 5.2.1 this implies that

$$\|x(t)\| \leq R\, \|x_0\|\, e^{(mR-\alpha)t}, \qquad 0 \leq t < t_1.$$

But x_0 and m are at our disposal, so we may choose m such that $mR < \alpha$ and $x(0) = x_0$, so that $\|x_0\| < d/2R$ implies $\|x(t)\| < d/2$ for $0 \leq t < t_1$.

Since $f(t, x)$ is defined for $\|x\| < a$ and $0 \leq t < \infty$, this implies that we can extend the solution $x(t)$, which exists locally at every point (t, x), $t > 0$, $\|x\| < a$, interval by interval, preserving the above bound. Hence, given any solution $x(t) = x(t; 0, x_0)$ with $\|x_0\| < d/2R$, it is defined on $0 \le t < \infty$ and satisfies $\|x(t; 0, x_0)\| < d/2$. But d can be made as small as desired, which implies that $x(t) \equiv 0$ is stable; and $mR < \alpha$ implies it is asymptotically stable.

From this result immediately follows the statement made concerning the nonlinear autonomous systems discussed in Section 4.4: the asymptotic stability or instability of trajectories of the linear system is preserved.

In this case we considered the system

$$\dot{x} = ax + by + \varepsilon_1(x, y), \qquad \dot{y} = cx + dy + \varepsilon_2(x, y), \tag{5}$$

where $ad - bc \neq 0$, $\varepsilon_i(x, y)$ were continuous together with their first partials and $\lim_{r \to 0} \varepsilon_i(x, y)/r = 0$.

Since the proof of Theorem 5.3.1 does not depend on the norm chosen, it follows that

(*a*) if the roots of the characteristic polynomial of $A = \begin{pmatrix} a & b \\ c & d \end{pmatrix}$ have negative real parts, then $(0, 0)$ is an asymptotically stable critical point of (5), or

(*b*) if the roots have positive real parts, then $(0, 0)$ is an unstable critical point of (5).

The last statement means that trajectories near $(0, 0)$ satisfy $\lim_{t \to -\infty} r(t) = 0$, and follows from the theorem by letting t approach $-\infty$ in the proof.

For the general system

$$\dot{x} = A(t)x + f(t, x), \tag{6}$$

with $A(t)$ a nonconstant matrix, the heuristic argument given at the beginning of this section fails. An example exists of a system for which $f(t, x)$ satisfies conditions (*i*) and (*ii*) above, and for which the solutions of $\dot{x} = A(t)x$ are asymptotically stable, but the solution $x(t) \equiv 0$ of (6) is unstable.

One reason for this deficiency is that the property of stability is rather delicate and may not be maintained under small changes on the right side of (6). A stronger definition of stability is the following.

DEFINITION: The solution $x(t; t_0, x_0) = x(t)$ is said to be *uniformly stable* if, given any $\varepsilon > 0$, there exists $\delta > 0$ such that any solution $x_1(t)$ satisfying $\|x(t_1) - x_1(t_1)\| < \delta$ for some $t_1 \geq t_0$ exists and satisfies $\|x(t) - x_1(t)\| < \varepsilon$ for $t \geq t_1$.

Note the distinction between stability and uniform stability. In the former a solution remains in an ε-neighborhood of $x(t; t_0, x_0)$ if it is close to the point x_0 at time t_0; other solutions may enter and leave the ε-neighborhood at later times. In the case of uniform stability, once a solution enters the ε-neighborhood of $x(t; t_0, x_0)$ it remains there. Briefly stated, in the definition of stability the number δ no longer depends on t_0.

Example: Consider the equation

$$\dot{x} = a(t)x, \qquad a(t) \text{ continuous on } 0 \leq t < \infty.$$

Then

$$x(t; t_1, x_1) = x_1 \exp\left[\int_{t_1}^{t} a(s)\, ds\right].$$

The solution $x(t) \equiv 0$ is uniformly stable if and only if the quantity

$$|x_1| \exp\left[\int_{t_1}^{t} a(s)\, ds\right]$$

can be made uniformly small for sufficiently small value of $|x_1|$. Therefore $x(t) \equiv 0$ is uniformly stable if and only if

$$\exp\left[\int_{t_1}^{t} a(s)\, ds\right]$$

is bounded above for $t \geq t_1 \geq 0$.

The conclusion in the last example also follows from the following result for linear systems. We assume that $A(t)$ is continuous for $t \geq t_0$.

LEMMA 5.3.1. *All solutions of $\dot{x} = A(t)x$ are uniformly stable if and only if there exists a positive constant M such that*

$$\|\Phi(t)\Phi^{-1}(s)\| < M, \qquad t_0 \leq s \leq t < \infty,$$

where $\Phi(t)$ is any fundamental matrix.

Proof. Given the solution $x(t) = x(t;\, t_0, x_0)$ and any fundamental matrix $\Phi(t)$, then for any $t_1 \geq t_0$ we have the expression $x(t) = \Phi(t)\,\Phi^{-1}(t_1)x(t_1)$. If $x_1(t) = \Phi(t)\Phi^{-1}(t_1)x_1(t_1)$ is any other solution, then given $\varepsilon > 0$ the relation

$$\|x(t_1) - x_1(t_1)\| < \varepsilon/M$$

implies

$$\|x(t) - x_1(t)\| \leq \|\Phi(t)\Phi^{-1}(t_1)\|\; \|x(t_1) - x_1(t_1)\| < \varepsilon$$

for $t \geq t_1 \geq t_0$, and hence $x(t)$ is uniformly stable.

To prove the converse, note that the hypotheses imply that the solution $x(t;\, t_0, 0) \equiv 0$ is uniformly stable, then proceed as in the proof of Theorem 5.2.1.

Finally we will prove a result for the system (6), in which we assume that the matrix $A(t)$ is continuous for $t \geq t_0$ and $f(t, x)$ satisfies

(*i*′) $f(t, x)$ is continuous for $\|x\| < a$, $t \geq t_0$, and
(*ii*′) there exists a continuous nonnegative function $\alpha(t)$ such that $\int_{t_0}^{\infty} \alpha(t)\, dt < \infty$ and

$$\|f(t, x)\| \leq \alpha(t)\, \|x\|.$$

Again note that (*ii*′) implies that $x(t) \equiv 0$, $t_0 \leq t < \infty$, is a solution of (6).

THEOREM 5.3.2. *Suppose that the solutions of $\dot{x} = A(t)x$ are uniformly or uniformly and asymptotically stable, and $f(t, x)$ satisfies conditions(i′) and (ii′) above. Then the solution $x(t) \equiv 0$ of the system*

$$\dot{x} = A(t)x + f(t, x)$$

is uniformly or uniformly and asymptotically stable.

Proof. Since all solutions of the linear system are uniformly stable, there exists by Lemma 5.3.1 a constant M such that for any fundamental matrix we have

$$\|\Phi(t)\Phi^{-1}(s)\| < M, \qquad t_0 \leq s \leq t < \infty.$$

If $t_1 \geq t_0$, then any solution $x(t)$ for which $\|x(t_1)\| < a$ satisfies the relation

$$x(t) = \Phi(t)\Phi^{-1}(t_1)x(t_1) + \int_{t_1}^{t} \Phi(t)\Phi^{-1}(s)f(s, x(s))\, ds,$$

for $t_1 < t < T$, where $\|x(t)\| < a$ for $t_1 \leq t < T$. Therefore

$$\|x(t)\| \leq M\,\|x(t_1)\| + M \int_{t_1}^{t} \alpha(s)\,\|x(s)\|\, ds,$$

and by Lemma 5.2.1 this implies

$$\|x(t)\| \leq M\,\|x(t_1)\| \exp\left[M \int_{t_1}^{t} \alpha(s)\, ds\right]$$

$$\leq M\,\|x(t_1)\| \exp\left[\int_{t_0}^{\infty} \alpha(s)\, ds\right] = K\,\|x(t_1)\|.$$

But given any $\varepsilon > 0$ with $\varepsilon < a$, then $\|x(t_1)\| < \varepsilon K^{-1}/2$ implies that $\|x(t)\| < \varepsilon/2$, and an argument similar to that given in Theorem 5.3.1 implies that $\|x(t)\| < \varepsilon/2$ for $t \geq t_1$. Therefore the solution $x(t) \equiv 0$ is uniformly stable.

Finally, if the solutions of the linear system are in addition asymptotically stable, then $\lim_{t\to\infty} \|\Phi(t)\| = 0$; hence, given any x_0 such that $\|x_0\| < a$ and any $\varepsilon > 0$, there exists a $T_0 > t_0$ such that $\|\Phi(t)x_0\| < \varepsilon$ for $t \geq T_0$. For the solution $x(t) = x(t;\, t_0, x_0)$ we then have for $t \geq T_0$

$$\|x(t)\| \leq \|\Phi(t)x_0\| + \int_{t_0}^{t} \|\Phi(t)\Phi^{-1}(s)\|\;\|f(s, x(s)\|\, ds$$

$$\leq \varepsilon + \int_{t_0}^{t} M\alpha(s)\,\|x(s)\|\, ds.$$

Again, by Lemma 5.2.1, this implies

$$\|x(t)\| \leq \varepsilon \exp\left[M \int_{t_0}^{\infty} \alpha(s)\, ds\right] = \varepsilon L, \qquad t \geq T_0,$$

and since ε was arbitrary and L does not depend on ε or T_0, we can conclude that $\lim_{t\to\infty} \|x(t)\| = 0$. Therefore the solution $x(t) \equiv 0$ is in addition asymptotically stable.

5.4 Liapunov's Direct Method

We will now briefly discuss an important method of studying the stability of solutions of the equation

$$\dot{x} = f(t, x), \tag{7}$$

where $x = x(t) = (x_1(t), \ldots, x_n(t))$ is an unknown vector function. The method is known as Liapunov's direct or second method and depends on being able to construct a particular type of function $V(t, x)$ from which the stability or instability of the solution in question can be determined.

We assume that $f(t, x) = (f_1(t, x), \ldots, f_n(t, x))$ satisfies the following conditions:

(*i*) $f(t, x)$ is defined and continuous in

$$\Gamma = \{(t, x) \mid \|x\| < a,\ r_1 < t < \infty\},$$

(*ii*) a condition assuring uniqueness of solutions $x(t; t_0, x_0)$ of (7) is satisfied at every point (t_0, x_0) in Γ, and

(*iii*) $f(t, 0) = 0$ for all t, and hence $x(t; t_0, 0) \equiv 0$ is a solution of (7) for $t_0 > r_1$.

For geometrical convenience by $\|x\|$ we will mean the Euclidean norm of x. Some preliminary definitions are now needed.

DEFINITION. The class K consists of all continuous, real-valued, strictly increasing functions $\varphi(r)$, $0 \le r \le a$, which vanish at $r = 0$.

Let $0 < b \le a$ and $t_0 > r_1$ and suppose that $V(t, x)$ is a real-valued function, continuous together with its first partial derivatives in the set

$$B = \{(t, x) \mid t_0 \le t < \infty,\ \|x\| \le b\}.$$

Furthermore, assume that $V(t, 0) = 0$ for $t \ge t_0$.

DEFINITION. The function $V(t, x)$ is positive (or negative) definite if there exists a function φ in the class K such that

$$V(t, x) \ge \varphi(\|x\|) \qquad \text{or} \qquad V(t, x) \le -\varphi(\|x\|)$$

for all (t, x) in B.

Examples

(*a*) The function

$$V(x, y) = x^4 + y^4$$

is positive definite, since $V(x, y) \geq \frac{1}{2}r^4$, where $r = \sqrt{x^2 + y^2}$.

(*b*) The function

$$V(t, x, y) = t(x^2 + y^2) - 2xy \cos t$$

is positive definite for $t \geq 2$, since $V(t, x, y) \geq r^2$.

DEFINITION. The function $V(t, x)$ is said to be descrescent or to admit an infinitesimal upper bound if there exists $h > 0$ and a function ψ in the class K such that

$$|V(t, x)| \leq \psi(\|x\|)$$

for $\|x\| < h$ and $t \geq t_0$.

Example: In the previous example both functions are descrescent with $\psi(r) = r^4$ and $\psi(r) = 3r^2$, respectively.

Now, given any function $V(t, x)$ as above and the equation (7), we denote by V' the function

$$V' = V'(t, x) = \sum_{i=1}^{n} \frac{\partial V}{\partial x_i} f_i(t, x) + \frac{\partial V}{\partial t}.$$

If $x = x(t)$, $t_0 \leq t \leq t_1$, is a solution of (7), then we can consider that $V(t, x(t)) = V(t)$. In this case $V' = \dot{V}$ is the derivative of V along the solution $x(t)$, and for simplicity we shall say that V' is the derivative of V.

THEOREM 5.4.1. *If a function $V(t, x)$ exists that is positive definite, and whose derivative V' is nonpositive, then the solution $x(t) \equiv 0$ of (7) is stable.*

Proof. Since V is positive definite, there exists a function φ in K such that $0 < \varphi(\|x\|) \leq V(t, x)$ for $0 < \|x\| \leq b$ and $t \geq t_0$. Given $\varepsilon > 0$, let

$$m_\varepsilon = \min_{\|x\| = \varepsilon} \varphi(\|x\|),$$

so $m_\varepsilon > 0$. Since V is continuous and $V(t, 0) = 0$, we can choose $\delta > 0$ so that $V(t_0, x_0) < m_\varepsilon$ if $\|x_0\| < \delta$. Furthermore, since V' is nonpositive, $t_1 \geq t_0$ and $\|x_0\| < \delta$ implies

$$\begin{aligned} V(t_1, x(t_1; t_0, x_0)) &\leq V(t_0, x(t_0; t_0, x_0)) \\ &= V(t_0, x_0) < m_\varepsilon. \end{aligned}$$

Now suppose that for some $t_1 > t_0$ we have $\|x(t_1; t_0, x_0)\| = \varepsilon$ when $\|x_0\| < \delta$. This would imply that

$$V(t_1, x(t_1; t_0, x_0)) \geq \varphi(\|x(t_1; t_0, x_0)\|) = \varphi(\varepsilon) \geq m_\varepsilon,$$

which is a contradiction. Therefore, if $\|x_0\| < \delta$, then the solution $x(t; t_0, x_0)$ is defined for $t \geq t_0$ and satisfies $\|x(t; t_0, x_0)\| < \varepsilon$, which implies that the solution $x(t) \equiv 0$ is stable.

To ensure that the zero solution of (7) is asymptotically stable, stricter conditions on the function V and its derivative are required.

THEOREM 5.4.2. *If a function $V(t, x)$ exists that is positive definite, descrescent, and whose derivative V' is negative definite, then the solution $x(t) \equiv 0$ of (7) is asymptotically stable.*

Proof. By the previous result the solution $x(t) \equiv 0$ is stable; therefore, given $\varepsilon > 0$, suppose there exists $\delta > 0$, $\lambda > 0$ and a solution $x(t; t_0, x_0)$ of (7) such that

$$\lambda \leq \|x(t; t_0, x_0)\| < \varepsilon, \qquad t \geq t_0, \quad \|x_0\| < \delta.$$

Since V' is negative definite, there exists a function γ in K such that

$$V'(t, x(t; t_0, x_0)) \leq -\gamma(\|x(t; t_0, x_0)\|).$$

Furthermore, since $\|x(t; t_0, x_0)\| \geq \lambda > 0$ for $t \geq t_0$, there exists a constant $d > 0$ such that

$$V'(t, x(t; t_0, x_0)) \leq -d < 0, \qquad t \geq t_0.$$

This implies that

$$\begin{aligned} V(t, x(t; t_0, x_0)) &= V(t_0, x_0) + \int_{t_0}^{t} V'\, dt \\ &\leq V(t_0, x_0) - d(t - t_0), \end{aligned}$$

and for sufficiently large t the right side will become negative, which contradicts V being positive definite. Hence no such λ exists and, since $V(t, x(t; t_0, x_0))$ is a positive decreasing function, it follows that $\lim_{t\to\infty} V(t, x(t; t_0, x_0)) = 0$. Therefore $\lim_{t\to\infty} \|x(t; t_0, x_0)\| = 0$ and this implies that the solution $x(t) \equiv 0$ is asymptotically stable.

DEFINITION. A function $V(t, x)$ satisfying the hypotheses of Theorem 5.4.1 is called a Liapunov function of the equation (7).

In the case of autonomous systems we may omit the dependence of V on t, and therefore delete the family K, which serves to ensure a uniformity with respect to t. Thus $V = V(x)$ is a positive (negative) definite function if

(*i*) V is continuous together with its partials in some neighborhood of the origin, and

(*ii*) $V(x) \geq 0$ (or ≤ 0), with equality only when $x = 0$.

For the case $n = 2$ the following geometric description is helpful.

For small positive c the curves $V(x, y) = c$ constitute a family of concentric loops enclosing the origin. The hypotheses of Theorem 5.4.1 imply that on small enough loops the direction of the vector field defined by the system $\dot{x} = P(x, y)$ $\dot{y} = Q(x, y)$ never points outward. Hence, once a trajectory of the equation is trapped inside such a loop, it cannot escape. The hypotheses of Theorem 5.4.2 imply that the vector field points inward.

Liapunov functions have been described for certain classes of differential equations, but how to proceed with any particular equation is partially a matter of experience and ingenuity. Fortunately Liapunov functions are often closely related to certain physical characteristics of the system described by the differential equation.

Examples

(*a*) Given the second-order equation

$$\ddot{x} + q(x) = 0,$$

where q is continuously differentiable, $q(0) = 0$ and $xq(x) > 0$, we consider the corresponding system

$$\dot{x} = y, \qquad \dot{y} = -q(x).$$

The hypotheses insure that (0, 0) is the only critical point. The total energy of the system is given by

$$V(x, y) = \frac{y^2}{2} + E(x) = \frac{y^2}{2} + \int_0^x q(s)\, ds,$$

where $E(x)$ is the potential energy integral.

The function V is continuously differentiable, $V(0, 0) = 0$, and $xq(x) > 0$ implies $V(x, y) > 0$ for $(x, y) \neq (0, 0)$ and therefore V is positive definite. Also,

$$V' = V_x \dot{x} + V_y \dot{y} = q(x)y + y(-q(x)) = 0,$$

and hence V is a Liapunov function and therefore the critical point (0, 0) is stable.

(*b*) For the system

$$\dot{x} = -y - x^3, \qquad \dot{y} = x - y^3,$$

we consider the function $V(x, y) = x^2 + y^2$. Certainly V is positive definite and, furthermore,

$$V' = 2x(-y - x^3) + 2y(x - y^3) = -2(x^4 + y^4)$$

is negative definite. Therefore the isolated critical point (0, 0) is asymptotically stable.

(*c*) Suppose we are given a set of functions $\{\varphi(x)\}$ called controls, such that for any φ the system

$$\dot{x} = f(x, \varphi(x))$$

has unique solutions, and $f(0, \varphi(0)) = 0$, so that the origin is a critical point. The object is to choose controls so that the trajectories of the corresponding system return to the origin—that is, to an equilibrium position.

A natural choice for a Liapunov function is $V(x) = d(x)$, the distance from x to the origin. The function $d(x)$ is positive definite, and if we wish to avoid having trajectories tend away from the origin we must choose $\varphi(x)$ so that $d'(x) < 0$ for $x \neq 0$. Finally, to ensure that the trajectories always return to a state of equilibrium, we must have $d'(0) = 0$; that is, $d'(x)$ must be negative definite.

Finally, we state the following instability theorem for the autonomous system

$$\dot{x} = f(x), \tag{8}$$

where $f(0) = 0$, and f is continuous together with its first partial derivatives in some neighborhood Γ of the origin. The functions $V(x)$ will as before be assumed to be continuous together with their first partial derivatives near the origin, say in Γ, and $V(0) = 0$.

THEOREM 5.4.3. *If there exists a function V such that V' is positive definite and in every neighborhood of the origin there is a point x_0 where $V(x_0) > 0$, then the solution $x(t) \equiv 0$ of (8) is unstable.*

Proof. Let $R > 0$ be sufficiently small, so that the ball

$$S(R) = \{x \mid \|x\| \leq R\}$$

lies in Γ. Let $M = \max_{\|x\| \leq R} V(x)$ and M is finite since V is continuous. Choose $r > 0$ so that $0 < r < R$ and by hypothesis there exists a point x_0 such that $0 < \|x_0\| < r$ and $V(x_0) > 0$. Along the trajectory $C: x = x(t; t_0, x_0), t \geq t_0$, V' is positive, and therefore $V(x(t; t_0, x_0))$, $t \geq t_0$, is an increasing function and $V(x(t_0; t_0, x_0)) > 0$. This implies that C cannot approach the origin. Furthermore, since V' is positive definite, the previous statement implies that

$$\inf_{t \geq t_0} V'(x(t; t_0, x_0)) = m > 0,$$

and therefore

$$V(x(t; t_0, x_0)) - V(x_0) \geq m(t - t_0)$$

for $t \geq t_0$. But the right side of the previous inequality can be made larger than M for t sufficiently large, which implies that C must leave the ball $S(R)$; therefore the origin is unstable.

Example: For the system

$$\dot{x} = 3x + y^2, \qquad \dot{y} = -2y + y^3,$$

consider the function $V(x, y) = x^2 - y^2$. Then V is continuous together with its first partials, $V(0, 0) = 0$, and V has positive values in any neighborhood of the origin. Furthermore,

$$\begin{aligned} V' &= 2x(3x + y^2) + (-2y)(-2y + x^3) \\ &= (6x^2 + 4y^2) + (2xy^2 - 2yx^3), \end{aligned}$$

and if $|x|$ and $|y|$ are sufficiently small the sign of V' is determined by the first term in parenthesis. Finally, $V'(0, 0) = 0$, and hence V' is positive definite near the origin; therefore the isolated critical point $(0, 0)$ is unstable.

5.5 Some Results for the Second-Order Linear Equation

In the study of differential equations, as in many other fields of mathematics, the study of specific equations can give much more information than is obtained from general theorems. This is certainly true for the equation

$$\ddot{y} + a(t)y = 0, \tag{9}$$

one of the most widely discussed equations in the mathematical literature. It should be noted that the general second-order linear equation

$$\ddot{z} + p(t)\dot{z} + q(t)z = 0$$

can be reduced to the form (9) by the transformation

$$z = y \exp\left(-\tfrac{1}{2}\int_0^t p(s)\,ds\right).$$

Here $y = y(t)$ is a scalar function, and we assume that $a(t)$ is real-valued and continuous on $0 \le t < \infty$. We will give a few results, which compare the behavior of solutions of (9) as t approaches ∞ with the solutions of a corresponding linear equation with constant coefficients.

THEOREM 5.5.1. *If*

$$\int_1^\infty t\,|a(t)|\,dt < \infty,$$

then $\lim_{t\to\infty} \dot{y}(t)$ *exists for any solution of* (9), *and any nontrivial solution is asymptotic to* $d_0 t + d_1$ *for some constants* d_0 *and* d_1 *not both zero.*

Proof. Let $y(t)$ be a solution of (9) satisfying $\dot{y}(t_0) = 1$, where $1 \le t_0 < \infty$. Integrating (9) twice from t_0 to t and using Dirichlet's formula (see Problem 6, Chapter 3), we obtain the relation

$$y(t) = c_1 + t - \int_{t_0}^{t} (t - s)a(s)y(s)\,ds,$$

where c_1 depends on t_0 and $y(t_0)$. Therefore, since $t \ge 1$, we obtain

$$|y(t)| \le (|c_1| + 1)t + t\int_{t_0}^{t} |a(s)|\,|y(s)|\,ds$$

or

$$\left|\frac{y(t)}{t}\right| \le (|c_1| + 1) + \int_{t_0}^{t} s\,|a(s)|\left|\frac{y(s)}{s}\right| ds$$

for $t \ge t_0$. Applying Lemma 5.2.1 we have

$$\left|\frac{y(t)}{t}\right| \le (|c_1| + 1)\exp\left[\int_{t_0}^{t} s\,|a(s)|\,ds\right]$$

$$\le (|c_1| + 1)\exp\left[\int_{t_0}^{\infty} s\,|a(s)|\,ds\right] = c_2 < \infty.$$

The constant c_2 depends on c_1, which is at our disposal, so we car choose t_0 such that

$$1 - c_2\int_{t_0}^{\infty} t\,|a(t)|\,dt > 0.$$

This implies that

$$\int_{t_0}^{t} |a(s)|\,|y(s)|\,ds \le c_2\int_{t_0}^{t} s\,|a(s)|\,ds < 1$$

for $t \geq t_0$. But

$$\dot{y}(t) = 1 - \int_{t_0}^{t} a(s)y(s)\,ds,$$

and therefore $\lim_{t\to\infty} \dot{y}(t)$ exists and is not zero. Therefore $y(t)$ is asymptotic to $d_0 t$, $d_0 \neq 0$, and now, using the result of Problem 6, Chapter 2, we can conclude that

$$u(t) = y(t) \int_{t}^{\infty} y^{-2}(s)\,ds$$

is another linearly independent solution asymptotic to $1/d_0$. This completes the proof.

Example: The Euler equation

$$\ddot{y} + mt^{-2}y = 0, \qquad m > \tfrac{1}{4},$$

has a fundamental system of solutions:

$$y_1(t) = t^{1/2}\cos(v \log t),$$
$$y_2(t) = t^{1/2}\sin(v \log t),$$

with $v > 0$. Therefore the assumption that $\lim_{t\to\infty} a(t) = 0$ is not sufficient to guarantee that solutions of (9) behave like the solution of $\ddot{y} = 0$.

We will now consider the case where solutions of (9) behave like those of the equation $\ddot{y} + y = 0$—that is, are bounded and oscillatory.

THEOREM 5.5.2 *If $\varphi(t)$ is continuously differentiable,*

$$\lim_{t\to\infty} \varphi(t) = 0,$$

and

$$\int_0^{\infty} |\dot{\varphi}(t)|\,dt < \infty,$$

then all solutions of the equation

$$\ddot{y} + (1 + \varphi(t))y = 0$$

are bounded.

Proof. Multiply the equation by $\dot{y}$, then integrate between 0 and t to obtain

$$\frac{(\dot{y}(t))^2}{2} + \frac{y^2(t)}{2} + \int_0^t \varphi(s)y(s)\dot{y}(s)\,ds = c_1,$$

for any solution $y(t)$ where c_1 is a constant. Now integration by parts and transposition lead to the relation

$$\begin{aligned} y^2(t)[1 + \varphi(t)] &= 2c_2 - (\dot{y}(t))^2 + \int_0^t \dot{\varphi}(s)y^2(s)\,ds \\ &\leq 2c_2 + \int_0^t \dot{\varphi}(s)y^2(s)\,ds, \end{aligned}$$

where c_2 is a constant. Since $\lim_{t\to\infty} \varphi(t) = 0$, we can choose t_0 large enough so that $1 + \varphi(t) \geq \frac{1}{2}$ for $t \geq t_0$, and therefore

$$|y^2(t)| \leq 4|c_2| + 2\int_0^t |\dot{\varphi}(s)|\,|y^2(s)|\,ds$$

for $t \geq t_0$. Therefore by Lemma 5.2.1 we have

$$\begin{aligned} |y^2(t)| &\leq 4|c_2| \exp\left[2\int_0^t |\dot{\varphi}(s)|\,ds\right] \\ &\leq 4|c_2| \exp\left[2\int_0^\infty |\dot{\varphi}(s)\,ds\right] = M < \infty, \end{aligned}$$

for $t \geq t_0$. Since $y(t)$ is continuous on $0 \leq t \leq t_0$, it follows that $y(t)$ is bounded.

To show that solutions of the last equation are oscillatory we need the following result, which is a form of the Sturm Comparison Theorem. By an *oscillatory solution* we mean one having an infinite number of zeros on $0 \leq t < \infty$.

THEOREM 5.5.3. *If all nontrivial solutions of* (9) *are oscillatory and if $b(t)$ is continuous and $b(t) \geq a(t)$, $t_0 \leq t < \infty$, then all nontrivial solutions of*

$$\ddot{x} + b(t)x = 0$$

are oscillatory.

Proof. From the two equations we obtain the relation

$$y\ddot{x} - x\ddot{y} + [b(t) - a(t)]xy = 0.$$

If t_1 and t_2 are two consecutive zeros of a nontrivial solution $y(t)$ of (9), we can assume that $t_0 \le t_1 < t_2$ and that $y(t) \ge 0$, $t_1 \le t \le t_2$. If $x(t)$ is any solution of the second equation, then integrating the previous expression from t_1 to t_2 gives

$$x(t_1)\dot{y}(t_1) - x(t_2)\dot{y}(t_2) + \int_{t_1}^{t_2} [b(t) - a(t)]x(t)y(t)\,dt = 0.$$

But if $x(t)$ did not change sign on $t_1 \le t \le t_2$, this would lead to a contradiction, since $\dot{y}(t_1) > 0$, $\dot{y}(t_2) < 0$ and $y(t)$ and $b(t) - a(t)$ are positive on $t_1 \le t \le t_2$. We may conclude that $x(t)$ has a zero in the interval $[t_1, t_2]$.

It follows that the nontrivial solutions of the equation $\ddot{y} + (1 + \varphi(t))y = 0$, where $\lim_{t\to\infty} \varphi(t) = 0$, are oscillatory. For t_0 sufficiently large we have $1 + \varphi(t) \ge \frac{1}{2}$ for $t \ge t_0$, and the solutions of $\ddot{y} + \frac{1}{2}y = 0$ are oscillatory, so we can apply Theorem 5.5.3.

From the above proof we can actually infer more: if the two solutions $x(t)$ and $y(t)$ have a common zero at $t = t_1$, then the solution $x(t)$ of $\ddot{x} + b(t)x = 0$ must have a zero in the interval $t_1 < t < t_2$. For, since $x(t_1) = 0$, we obtain the relation

$$x(t_2)\dot{y}(t_2) = \int_{t_1}^{t_2} [b(t) - a(t)]x(t)y(t)\,dt.$$

The right side is positive, so if $x(t)$ were positive in $t_1 < t < t_2$ we would have a contradiction, since $\dot{y}(t_2) < 0$.

We conclude the discussion of the equation (9) by considering the case where $a(t)$ is a strictly increasing unbounded function and we will assume that $a(t)$ is continuously differentiable.

THEOREM 5.5.4. *If* $\lim_{t\to\infty} a(t) = \infty$ *monotonically, then all solutions of* (9) *are bounded.*

Proof. Multiplying the equation by $\dot{y}$, integrating between 0 and t, and then integrating by parts lead to the relation

$$\frac{(\dot{y}(t))^2}{2} + a(t)\frac{y^2(t)}{2} - \int_0^t \frac{y^2(s)}{2}\dot{a}(s)\,ds = c_1$$

for any solution $y(t)$ of (9) where c_1 is a constant. We may assume that $a(t) > 0$ for $t > 0$ and therefore, since $\dot{a}(t) \geq 0$, we have

$$\left|\frac{a(t)y^2(t)}{2}\right| \leq |c_1| + \int_0^t \left|\frac{a(s)y^2(s)}{2}\right|\left|\frac{\dot{a}(s)}{a(s)}\right| ds.$$

Applying Lemma 5.2.1 then gives

$$\left|\frac{a(t)y^2(t)}{2}\right| \leq |c_1| \exp\left[\int_0^t \left|\frac{\dot{a}(s)}{a(s)}\right| ds\right],$$

which implies that

$$|y^2(t)| \leq 2|c_1|\,|a(0)|^{-1}, \qquad t \geq 0,$$

so all solutions are bounded.

If $\lim_{t\to\infty} a(t) = \infty$ monotonically, then certainly $a(t) > \varepsilon > 0$ for all t greater than some t_0, and we can conclude that the solutions of (9) are oscillatory. The amplitude of the oscillation never increases, as the following theorem shows. Note that between every two zeros of a solution will occur a zero of its derivative by Rolle's Theorem.

THEOREM 5.5.5. *Suppose that* $a(t)$ *is continuously differentiable and* $a(t) > 0$, $\dot{a}(t) \geq 0$ *on* $0 \leq t < \infty$. *If* $y(t)$ *is a solution of* (9) *and* t_1 *and* t_2 *are two consecutive zeros of its derivative, then* $|y(t_2)| \leq |y(t_1)|$.

Proof. We may assume $\dot{a}(t) \not\equiv 0$, $t_1 \leq t \leq t_2$; hence multiplying by $2\dot{y}(t)$ and integrating from t_1 to t_2 gives

$$\dot{y}^2(t)\Big]_{t_1}^{t_2} + 2\int_{t_1}^{t_2} a(t)y(t)\dot{y}(t)\,dt = 0.$$

By hypothesis the first expression is zero, and by a further integration by parts we have

$$a(t_2)y^2(t_2) - a(t_1)y^2(t_1) = \int_{t_1}^{t_2} \dot{a}(t)y^2(t)\,dt.$$

Since $\dot{y}(t)$ does not change sign in $t_1 \le t \le t_2$, the solution $y(t)$ is strictly monotonic in that interval. If $y^2(t_2) > y^2(t_1)$, then

$$\int_{t_1}^{t_2} \dot{a}(t)y^2(t)\,dt < [a(t_2) - a(t_1)]y^2(t_2),$$

which implies that

$$a(t_1)[y^2(t_2) - y^2(t_1)] < 0.$$

This is a contradiction, and we conclude that $y^2(t_2) \le y^2(t_1)$, which gives the desired result.

Problems

1. Given the nonhomogeneous system

$$\dot{x} = A(t)x + B(t),$$

where $A(t)$ amd $B(t)$ are continuous on $t_0 \le t \le \infty$, prove that

(*i*) if all solutions are bounded, then they are stable, and

(*ii*) if all solutions are stable and one is bounded, then all solutions are bounded.

2. Let $\Phi(t)$ be a fundamental matrix of the system $\dot{x} = A(t)x$, where $A(t)$ is real-valued and continuous on $0 \le t < \infty$.

(a) Show that the transpose of $\Phi^{-1}(t)$ satisfies the matrix differential equation

$$\dot{X} = -A(t)^{\mathrm{T}}X,$$

where $A(t)^{\mathrm{T}}$ is the transpose of $A(t)$.

(b) Show that if the solutions of the system $\dot{x} = A(t)x$ are bounded and

$$\liminf_{t\to\infty} \int_0^t \operatorname{tr} A(s)\,ds > -\infty,$$

then $\|\Phi^{-1}(t)\|$ is bounded. (Hint: Express $\Phi^{-1}(t)$ in terms of the adjoint matrix of $\Phi(t)$.)

3. Use the results of the previous problem to show that if the solutions of $\dot{x} = A(t)x$ are bounded, and

(*i*) the matrix $C(t)$ is continuous on $0 \leq t < \infty$ and $\int_0^\infty \|C(t)\|\, dt < \infty$

and

(*ii*) $\liminf_{t\to\infty} \int_0^\infty \operatorname{tr} A(s)\, ds > -\infty$,

then the solutions of the system

$x = [A(t) + C(t)]x$

are bounded and hence stable.

4. The system $\dot{x} = A(t)x$, where $A(t)$ is real-valued and continuous on $0 \leq t < \infty$, is said to be stable if all its solutions are stable. It is said to be *restrictively stable* if it and its adjoint system $x = -A(t)^{\mathsf{T}}x$ are stable. Prove the following.

(a) A necessary and sufficient condition for restrictive stability is that there exist a constant M such that

$$\|\Phi(t)\Phi^{-1}(s)\| \leq M, \qquad t \geq 0,\ s \geq 0,$$

where $\Phi(t)$ is the fundamental matrix satisfying $\Phi(0) = I$.

(b) If the system is stable and

$$\liminf_{t\to\infty} \int_0^t \operatorname{tr} A(s)\, ds > -\infty,$$

then it is restrictively stable.

(c) If the adjoint system is stable and

$$\limsup_{t\to\infty} \int_0^t \operatorname{tr} A(s)\, ds < \infty,$$

then the system is restrictively stable.

5. Let $A(t)$ be a continuous real-valued square matrix on $0 \leq t < \infty$. Demonstrate the following.

(a) Every solution of the system $\dot{x} = A(t)x$ satisfies the relation

$$\|x(t)\| \leq \|x(0)\| \exp\left[\int_0^t \|A(s)\|\, ds\right].$$

(b) Use the previous result to show that if $\int_0^\infty \|A(t)\|\,dt < \infty$, then every solution $x(t)$ of the system has a finite limit as t approaches ∞. (Hint: Show that $A(t)x(t)$ is integrable on $0 \le t < \infty$.)

6. The system $\dot{x} = A(t)x$ is said to possess *linear asymptotic equilibrium* if, given any vector c, there exists a solution $x(t)$ satisfying $\lim_{t\to\infty} x(t) = c$.

Using 5(b) above, show that the condition $\int_0^\infty \|A(t)\|\,dt < \infty$ implies that the system has linear asymptotic equilibrium.

7. Using appropriate Liapunov functions, determine the type of stability or the instability of the following systems.

(a) $\dot{x} = -x + y + xy,$

$\dot{y} = x - y - x^2 - y^3.$

(b) $\dot{x} = x - 3y + x^3,$

$\dot{y} = -x + y - y^2.$

(c) $\dot{x} = -x - 2y + xy^2,$

$\dot{y} = 3x - 3y + y^3.$

(d) $\dot{x} = -4y \pm xe^{x+y}$

$\dot{y} = 4x \pm ye^{x+y}$

8. Let $f(x)$ and $g(x)$ be even and odd polynomials respectively and consider the second-order equation

$$\ddot{x} + f(x)\dot{x} + g(x) = 0.$$

(a) Show that the equation is equivalent to the system

$$\dot{x} = y - F(x), \qquad \dot{y} = -g(x),$$

where $F(x) = \int_0^x f(s)\,ds$.

(b) Let $G(x) = \int_0^x g(s)\,ds$ and suppose there exist positive constants a and b such that $g(x)F(x) > 0$ for $0 \le |x| < a$, and $G(x) < b$ implies $|x| < a$. Show that this implies

$$V(x, y) = \frac{y^2}{2} + G(x)$$

is a Liapunov function for the above system in the region $|x| < a$, $y^2 < 2b$, and that $(0, 0)$ is an asymptotically stable critical point.

9. Apply the previous results to the van der Pol equation

$$\ddot{x} + \varepsilon(x^2 - 1)\dot{x} + x = 0,$$

to show that (0, 0) is an asymptotically stable critical point if $\varepsilon < 0$. Find the values of the constants a and b.

10. Consider the second-order linear equation

$$\ddot{y} + p(t)\dot{y} + q(t)y = 0,$$

where $p(t)$ and $q(t)$ are real-valued and continuous on $r_1 < t < r_2$. Show that if $y_1(t)$ and $y_2(t)$ are a real-valued fundamental pair of solutions, then $y_1(t)$ must vanish between any two consecutive zeros of $y_2(t)$ (Sturm Separation Theorem).

11. Estimate the number of zeros of any nontrivial solution of the equation

$$\ddot{y} + ky = 0, \qquad k \text{ a positive constant,}$$

that can be contained in the interval $a \le t \le b$. Use this result and Theorem 5.5.3 to estimate the number of zeros of a nontrivial solution of the given equation in the interval indicated.

(a) $\ddot{y} + 5ty = 0, \quad 1 \le t \le 10.$

(b) $\ddot{y} + t^{-2}y = 0, \quad 1 \le t \le 10.$

12. Given the equation of Problem 10 above, show that a nontrivial solution cannot have an infinite number of zeros on any closed interval $a \le t \le b$ contained in $r_1 < t < r_2$.

CHAPTER

6

Existence, Uniqueness, and Related Topics

6.1 Proof of the Existence and Uniqueness of Solutions

In Section 1.3 we stated without proof Theorem 1.3.1, a theorem giving conditions under which the existence and uniqueness of solutions of an ordinary differential equation are assured. The proof will be given in this section, and depends on the *method of successive approximations*.

This method, which is frequently used in many different mathematical settings to prove existence of solutions, may be described as follows: we choose an initial approximation $x_0(t)$ to a solution, based on the initial data. Furthermore, an algorithm is devised whereby we can construct successive approximations $x_1(t), x_2(t), \ldots, x_n(t), \ldots$ to a solution. Finally, we show that the sequence $\{(x_n(t)\}$ converges in some suitable topology to a solution $x(t)$.

Example: Given the differential equation

$$\dot{x} = x, \qquad x(0) = 1,$$

where $x = x(t)$ is a scalar function, then any solution must satisfy the relation

$$x(t) = 1 + \int_0^t x(s)\, ds.$$

Therefore, given $x_n(t)$, an approximation to a solution $x(t)$, we construct $x_{n+1}(t)$ by the relation

$$x_{n+1}(t) = 1 + \int_0^t x_n(s)\, ds.$$

As a first approximation, let $x_0(t) \equiv 1$, and we easily verify that

$$x_1(t) = 1 + t, \qquad x_2(t) = 1 + t + t^2/2, \ldots,$$

and, in general,

$$x_n(t) = \sum_{k=0}^{n} \frac{t^k}{k!},$$

the nth partial sum of the Taylor expansion of the solution $x(t) = e^t$.

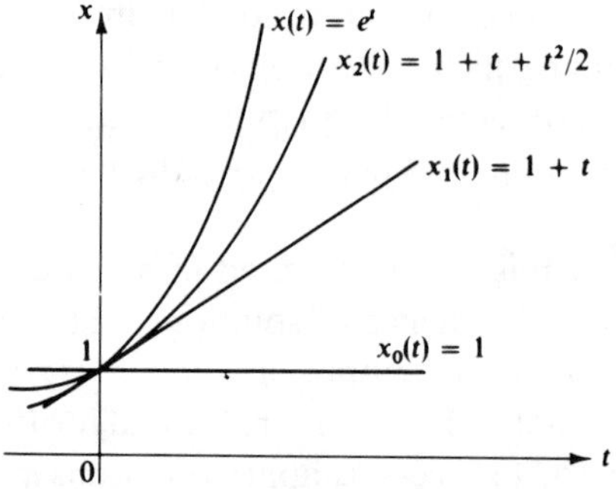

To apply the method of successive approximations in the proof of Theorem 1.3.1 we will need some additional terminology.

Given a positive integer n and any closed interval $a \le t \le b$, we denote by $C[a, b]$ the set of all n-dimensional vector functions $x(t) = (x_1(t), \ldots, x_n(t))$, which are continuous on $a \le t \le b$. We introduce the following norm in $C[a, b]$:

$$\|x\|_C = \max_{a \le t \le b} \|x(t)\| = \max_{a \le t \le b} \sum_{j=1}^{n} |x_j(t)|.$$

It is easily verified that $C[a, b]$ becomes a metric space with the metric given by

$$\operatorname{dist}(x, y) = \|x - y\|_C,$$

where $x = x(t)$ and $y = y(t)$ are in $C[a, b]$.

Furthermore, note that given any Cauchy sequence $\{x_s\}$ in $C[a, b]$—that is, a sequence of elements having the property that

$$\lim_{r,s\to\infty} \|x_r - x_s\|_C = \lim_{r,s\to\infty} \max_{a\le t\le b} \|x_r(t) - x_s(t)\| = 0,$$

then, by the choice of norm, this is equivalent to uniform convergence on $a \le t \le b$ of the sequence $\{x_s\}$. Since the limit of a uniformly convergent sequence of continuous functions is a continuous function, we may conclude that $C[a, b]$ is a *complete metric space*.

Let the constant $c > 0$ and the point x_0 in R^n be given, and suppose $f(t, x)$ is an n-dimensional vector-valued function that is continuous on the open set

$$E = \{(t, x) \mid a < t < b, \|x - x_0\| < c\}.$$

Let $x = x(t)$ belong to $C[a, b]$ and suppose that the point $(t, x(t))$ belongs to E for $r_1 \le t \le r_2$. Then, given any t_0 such that $r_1 < t_0 < r_2$, the mapping

$$\begin{aligned}\mathscr{A}x &= y = y(t)\\ &= x_0 + \int_{t_0}^{t} f(s, x(s))\, ds, \qquad r_1 \le t \le r_2,\end{aligned}$$

defines a continuous function on $r_1 \le t \le r_2$. Since $x = x(t)$ also belongs to $C[r_1, r_2]$, we may conclude that $\mathscr{A}$ is a map from $C[r_1, r_2]$ into itself. Since $y(t_0) = x_0$, it follows that for all t in some neighborhood of t_0 the point $(t, y(t))$ belongs to E.

Finally, if we consider the differential equation

$$\dot{x} = f(t, x), \tag{1}$$

then any solution $x = x(t)$ that satisfies the initial condition $x(t_0) = x_0$ and is defined on $r_1 \le t \le r_2$ also satisfies the relation

$$x(t) = x_0 + \int_{t_0}^{t} f(s, x(s))\, ds, \qquad r_1 \le t \le r_2.$$

But the right side of the last expression is $\mathscr{A}x$, so we can conclude that any solution $x = x(t)$, $r_1 \le t \le r_2$, of the differential equation (1) satisfies the relation

$$x = \mathscr{A}x. \tag{2}$$

Since x certainly belongs to $C[r_1, r_2]$, the last relation asserts that a solution $x = x(t), r_1 \le t \le r_2$, of (1) is a *fixed point* of the mapping $\mathscr{A}$.

For the reader's benefit we restate Theorem 1.3.1 and then proceed to its proof.

THEOREM. *Let the equation* $(*)$ $\dot{x} = f(t, x)$ *be given, where* $f(t, x)$ *is defined and continuous in some domain* B *contained in* R^{n+1} *and furthermore suppose that* $\partial f/\partial x_i$, $i = 1, \ldots, n$ *are defined and continuous in* B. *Then for every point* (t_0, x_0) *in* B *there exists a unique solution* $x = x(t)$ *of* $(*)$ *satisfying* $x(t_0) = x_0$ *and defined in some neighborhood of* (t_0, x_0).

Proof. The domain B is an open connected set (assumed to be nonempty), and since (t_0, x_0) belongs to B there exist positive numbers a and b such that the closed bounded set

$$\Gamma = \{(t, x) \,|\, |t - t_0| \le a, \ \|x - x_0\| \le b\}$$

is contained in B. The functions f and $\partial f/\partial x_i$, $i = 1, \ldots, n$, are continuous in B, and therefore continuous in Γ. This implies that there exist positive numbers m and k such that

$$\|f(t, x)\| \le m, \qquad \left|\frac{\partial f_i(t, x)}{\partial x_j}\right| \le k, \qquad i, j = 1, \ldots, n,$$

whenever (t, x) is in Γ. From the last inequality and the mean value theorem it follows that

$$\|f(t, x_1) - f(t, x_2)\| \le nk \, \|x_1 - x_2\|$$

whenever (t, x_1) and (t, x_2) belong to Γ, since Γ is convex.

Now choose $r > 0$ such that

$$(i) \ r \le a, \qquad (ii) \ r \le b/m, \qquad (iii) \ r < 1/nk,$$

and let Γ_r be the closed bounded subset of Γ defined by

$$\Gamma_r = \{(t, x) \,|\, |t - t_0| \le r, \ \|x - x_0\| \le b\}.$$

The reasons for imposing the restrictions (*i*), (*ii*), and (*iii*) on r will be clear shortly.

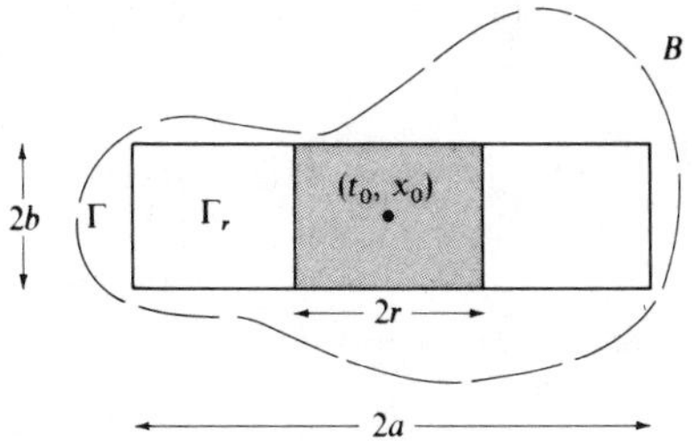

Let $\mathscr{C}_r$ be the set of all functions $x(t) = (x_1(t), \ldots, x_n(t))$ satisfying the following conditions:

(*a*) the function $x(t)$ is continuous for $|t - t_0| \leq r$, and

(*b*) the point $(t, x(t))$ belongs to Γ_r for $|t - t_0| \leq r$.

The conditions imply that $\|x(t) - x_0\| \leq b$ for $|t - t_0| \leq r$, and therefore $\mathscr{C}_r$ is that subset of functions belonging to $C[t_0 - r, t_0 + r]$, whose graphs lie completely in Γ_r.

For $x = x(t)$ in $\mathscr{C}_r$ let $y = y(t)$ be defined by

$$y = y(t) = \mathscr{A}x = x_0 + \int_{t_0}^{t} f(s, x(s))\, ds, \; |t - t_0| \leq r.$$

Then

$$\|y(t) - x_0\| = \left\| \int_{t_0}^{t} f(s, x(s))\, ds \right\|$$
$$\leq \int_{t_0}^{t} \|f(s, x(s))\|\, ds \leq m\,|t - t_0| \leq mr \leq b,$$

and we see that condition (*ii*) on the choice of r ensures that the graph of $y = y(t) = \mathscr{A}x$ is in Γ_r for any x in $\mathscr{C}_r$. Since $y(t)$ is a continuous function, it follows that the choice of r ensures that $\mathscr{A}$ is a map from $\mathscr{C}_r$ into itself. Furthermore, note that $y(t_0) = x_0$, and hence y satisfies the initial conditions.

Suppose that $x_1 = x_1(t)$ and $x_2 = x_2(t)$ belong to $\mathscr{C}_r$. Then

$$\begin{aligned}\|\mathscr{A}x_1 - \mathscr{A}x_2\| &= \left\|\int_{t_0}^{t} [f(s, x_1(s)) - f(s, x_2(s))]\, ds\right\| \\ &\le \int_{t_0}^{t} \|f(s, x_1(s)) - f(s, x_2(s))\|\, ds \\ &\le nk \int_{t_0}^{t} \|x_1(s) - x_2(s)\|\, ds \\ &\le nkr \left[\max_{|t-t_0|\le r} \|x_1(t) - x_2(t)\|\right] = \alpha \|x_1 - x_2\|_C .\end{aligned}$$

Thus condition (*iii*) on the choice of r ensures that $\alpha = nkr < 1$. It follows that

$$\begin{aligned}\max_{|t-t_0|\le r} \|\mathscr{A}x_1 - \mathscr{A}x_2\| &= \|\mathscr{A}x_1 - \mathscr{A}x_2\|_C \\ &\le \alpha \|x_1 - x_2\|_C\end{aligned}$$

for any x_1 and x_2 in $\mathscr{C}_r$, where $0 < \alpha < 1$. Since $\| \ \|_C$ is a measure of the distance in $\mathscr{C}_r$, the last relation implies that the distance between the images (under $\mathscr{A}$) of two "points" in $\mathscr{C}_r$ is less than the distance between them; that is, $\mathscr{A}$ is *contraction mapping* from $\mathscr{C}_r$ into itself.

We commence the process of successive approximations by letting $x_0(t) \equiv x_0$, $|t - t_0| \le r$, be the first approximation. Given no other information it is probably a poor but logical choice for a first approximation. We then define

$$x_1 = x_1(t) = \mathscr{A}x_0,$$

$$x_2 = x_2(t) = \mathscr{A}x_1, \ldots, x_k = x_k(t) = \mathscr{A}x_{k-1}, \ldots,$$

and from our definition of the mapping $\mathscr{A}$, $x_j(t_0) = x_0$, and $x_j(t)$ belong to $\mathscr{C}_r$, $j = 0, 1, \ldots, k, \ldots$. Note also that

$$x_k = \mathscr{A}x_{k-1} = \mathscr{A}(\mathscr{A}x_{k-2}) = \mathscr{A}^2 x_{k-1} = \cdots = \mathscr{A}^k x_0$$

for any k, where by $\mathscr{A}^k$ we mean the mapping $\mathscr{A}$ applied k times.

Let s and m be positive integers and $s > m$. Then by the contraction property, for the approximations x_s and x_m we have

$$\begin{aligned}\|x_m - x_s\|_C &= \|\mathscr{A}x_{m-1} - \mathscr{A}x_{s-1}\|_C \le \alpha \|x_{m-1} - x_{s-1}\|_C \\ &= \alpha \|\mathscr{A}x_{m-2} - \mathscr{A}x_{s-2}\|_C \le \alpha^2 \|x_{m-2} - x_{s-2}\|_C \le \cdots \\ &\le \alpha^m \|x_0 - x_{s-m}\|_C .\end{aligned}$$

By the triangle inequality we have

$$\begin{aligned}\alpha^m \|x_0 - x_{s-m}\|_C \le \alpha^m \{&\|x_0 - x_1\|_C \\ &+ \|x_1 - x_2\|_C + \cdots + \|x_{s-m-1} - x_{s-m}\|\},\end{aligned}$$

and therefore

$$\begin{aligned}\alpha^m \|x_0 - x_{s-m}\|_C &\le \alpha^m \|x_0 - x_1\|_C\{1 + \alpha + \alpha^2 + \cdots + \alpha^{s-m-1}\} \\ &\le \frac{\alpha^m}{1-\alpha}\|x_0 - x_1\|_C \le \frac{\alpha^m b}{1-\alpha}.\end{aligned}$$

Combining all the above, for any positive integers s and m, $s > m$, the successive approximations x_s and x_m satisfy

$$\|x_m - x_s\|_C \le \frac{\alpha^m b}{1-\alpha}, \qquad 0 < \alpha < 1.$$

But the right side approaches zero as m approaches infinity, which implies that the sequence $x_j = x_j(t)$, $j = 0, 1, 2, \ldots,$ of successive approximations is a Cauchy sequence.

Since $C[t_0 - r, t_0 + r]$ is a complete metric space, it follows that there exists a continuous function $x = x(t)$ defined on $|t - t_0| \le r$ such that

(*a*) $\lim_{j\to\infty} x_j(t) = x(t)$ uniformly on $|t - t_0| \le r$,

(*b*) the point $(t, x(t))$ is contained in Γ_r for $|t - t_0| \le r$ and $x(t_0) = x_0$.

Hence $x = x(t)$ belongs to $\mathscr{C}_r$, and therefore for any j

$$\|\mathscr{A}x - \mathscr{A}x_j\|_C \le \alpha \|x - x_j\|_C .$$

But the right side approaches zero as j approaches ∞, which implies that

$$\mathscr{A}x = \lim_{j\to\infty} \mathscr{A}x_j = \lim_{j\to\infty} x_{j+1} = x.$$

Therefore $x = x(t)$ is a fixed point of the mapping $\mathscr{A}$ and so is a solution of the differential equation satisfying $x(t_0) = x_0$. Existence of a solution is proved.

To prove uniqueness, suppose there exists another solution $y = y(t)$ satisfying $y(t_0) = x_0$ and defined on $|t - t_0| \le r_1$. Then there exists a positive number $s \le \min(r, r_1)$ such that $(t, y(t))$ lies in Γ_s for $|t - t_0| \le s$. Since $x = x(t)$ and $y = y(t)$ are solutions, we have

$$\max_{|t-t_0|\le s} \|x(t) - y(t)\| = \|x - y\|_C = \|\mathscr{A}x - \mathscr{A}y\|_C$$
$$\le \alpha \|x - y\|_C, \qquad 0 < \alpha < 1.$$

This can hold only if $\|x - y\|_C = 0$, which implies that $x(t) = y(t)$, $|t - t_0| \le s$; hence uniqueness is proved, and this completes the proof of the theorem.

Example: Consider the equation

$$\dot{x} = x^2 + t^2, \qquad x(0) = 1,$$

and therefore $f(t, x) = x^2 + t^2$, $(t_0, x_0) = (0, 1)$. Since B is the entire (t, x)-plane, let us choose as Γ the unit square

$$\Gamma = \{(t, x) \,|\, |t| \le 1, |x - 1| \le 1\}.$$

Therefore $a = b = 1$ and

$$|f(t, x)| = |x^2 + t^2| \le 5, \qquad \left|\frac{\partial f}{\partial x}\right| = |2x| \le 4$$

for (t, x) in Γ, and hence $m = 5$, $k = 4$. From the conditions (*i*), (*ii*), and (*iii*) on r we deduce that $0 < r \le \frac{1}{5}$.

The mapping $\mathscr{A}$ is given by

$$y(t) = \mathscr{A}x = 1 + \int_0^t [x^2(s) + s^2]\, ds, \qquad |t| \le \tfrac{1}{5},$$

and, letting the first approximation be $x_0(t) = x_0 = 1$, we have

$$x_1(t) = \mathscr{A}x_0 = 1 + \int_0^t (1 + s^2)\,ds$$
$$= 1 + t + \frac{t^3}{3},$$

$$x_2(t) = \mathscr{A}x_1 = 1 + \int_0^t \left[\left(1 + s + \frac{s^3}{3}\right)^2 + s^2\right] ds$$
$$= 1 + t + t^2 + \frac{2}{3}t^3 + \frac{1}{6}t^4 + \frac{2}{15}t^5 + \frac{1}{63}t^7,$$

and so forth. The successive approximations generated in this manner will converge uniformly to a solution $x(t)$ satisfying $x(0) = 1$ and defined for $|t| \leq \frac{1}{5}$.

It should be noted that because of the restrictions on r, the existence theorem is *local* in nature. The successive approximations converge to a solution satisfying the initial conditions and defined *only* in a neighborhood of t_0, even in the case where B is of the form

$$B = \{(t, x) \mid -\infty < t < \infty,\ \|x - x_0\| < b\},$$

or B is all of R^{n+1}.

Whether the solution so obtained is actually defined for other t is another question. In the example given at the beginning of this chapter the successive approximations converged to a solution defined for $-\infty < t < \infty$. For the equation $\dot{x} = 1 + x^2$ the solutions are of the form $x(t) = \tan(t - h)$, so the successive approximations would converge to a solution defined in an interval of length less than π. In both cases $B = R^2$.

In the proof of the existence theorem the continuity in B of the partial derivatives $\partial f_i/\partial x_j$ was not explicitly used. As the proof indicates we could substitute the following Lipschitz condition.

There exists a constant $k > 0$ such that

$$\|f(t, x_1) - f(t, x_2)\| \leq k\,\|x_1 - x_2\|$$

for every pair of points (t, x_1) and (t, x_2) in B.

This leads to the following result.

THEOREM 6.1.2. *Let the equation $\dot{x} = f(t, x)$ be given where $f(t, x)$ is defined and continuous in some domain B contained in R^{n+1}. If the above Lipschitz condition is satisfied, then corresponding to every point (t_0, x_0) in B there exists a unique solution $x(t)$ of the equation satisfying $x(t_0) = x_0$ and defined in some neighborhood of (t_0, x_0).*

The above theorem is actually a stronger result, since there exist continuous functions satisfying a Lipschitz condition but not having continuous first partial derivatives, for example, $f(x) = |x|$ in any neigborhood of $x = 0$. However, in many cases it is easier to verify the continuity of the first partial derivatives than to determine whether a Lipschitz condition is satisfied. Furthermore, note that the Lipschitz condition need only be satisfied in some neighborhood of the point (t_0, x_0) and the value of the constant k can vary with the choice of the domain B.

Finally, it should be remarked that the existence theorem is an application of a more general theorem, which states that a contraction mapping from a complete metric space into itself has a fixed point. However, although this principle is applicable in many different instances, it may turn out badly as an approximation technique. In the case of ordinary differential equations the convergence may be too slow, or the successive integrations may become cumbersome. If we are interested in finding the values of a solution near the initial point $x(t_0)$, we should preferably use a more refined numerical technique.

6.2 Continuation of Solutions and the Maximum Interval of Existence

In this section we will first discuss the question mentioned in the previous section—whether the solutions $x = x(t)$ of the equation

$$\dot{x} = f(t, x), \tag{3}$$

which are uniquely defined in the neighborhood of an initial point, can be extended. We will also discuss the maximum interval of existence for linear equations.

Recall that given a point (t_0, x_0) in the domain B, we were able to find a positive number r and a unique solution $x = x(t)$ of (3) such that $x(t_0) = x_0$, and $x(t)$ was defined for $|t - t_0| \le r$. Certainly the point $(t_0 + r,\ x(t_0 + r))$ is in B (since B is an open connected set) and we can then repeat the method of successive approximations to obtain a solution defined in a neighborhood of the point $(t_0 + r, x(t_0 + r))$.

This solution would then be defined on an interval $t_0 - r \le t \le t_0 + r + \alpha$, $\alpha > 0$, and since it must agree with $x(t)$ for $|t - t_0| \le r$, we are justified in calling it the *continuation* (to the right) of $x(t)$ to the larger interval. Proceeding in this way we would eventually (since B is open) obtain an open interval $r_1 < t < r_2$ and a solution $x = x(t)$, defined on $r_1 < t < r_2$ and such that $x(t_0) = x_0$.

This implies that the points $(t, x(t))$, $r_1 < t < r_2$, are in B, and we will assume that r_1 and r_2 are finite. We would expect that if the points $(r_1, x(r_1 + 0))$ or $(r_2, x(r_2 - 0))$ were in the domain B, then under suitable conditions we should be able to extend the solution to a larger interval. By $x(r_1 + 0)$ we mean the limit of $x(t)$ as t approaches r_1 from the right and $x(r_2 - 0)$ is the limit as t approaches r_2 from the left.

THEOREM 6.2.1. *Let $f(t, x)$ satisfy the hypotheses of Theorem 1.3.1 in some domain B contained in R^{n+1}, and suppose $x = x(t)$, $r_1 < t < r_2$, is a solution of (3). If f is bounded on B, then*

$$\lim_{t \to r_1 + 0} x(t) \qquad \text{and} \qquad \lim_{t \to r_2 - 0} x(t)$$

both exist, and furthermore if $(r_1, x(r_1 + 0))$ or $(r_2, x(r_2 - 0))$ are in B, the solution $x(t)$ can be continued to the left or right.

Proof. We may assume the solution satisfies $x(t_0) = x_0$ for some (t_0, x_0) in B, $r_1 < t_0 < r_2$, and therefore $x = x(t)$ satisfies

$$x(t) = x_0 + \int_{t_0}^{t} f(s, x(s))\, ds, \qquad r_1 < t < r_2.$$

If $b_n = x(r_2 - 1/n)$, then for n sufficiently large and $m > n$ we have

$$\|b_m - b_n\| \le \int_{r_2 - 1/n}^{r_2 - 1/m} \|f(s, x(s))\|\, dx \le M \left| \frac{1}{m} - \frac{1}{n} \right|,$$

where M is a bound for $f(t, x)$ in B. This implies that $\{b_n\}$ is a Cauchy sequence, from which it follows that $\lim_{t \to r_2 - 0} x(t)$ exists, and similarly for $\lim_{t \to r_1 + 0} x(t)$.

Suppose that the point $(r_2, x(r_2 - 0))$ is in B; then the function $\hat{x}(t)$ defined by

$$\hat{x}(t) = x(t), \qquad r_1 < t < r_2, \qquad \hat{x}(r_2) = x(r_2 - 0))$$

is a solution of (3), defined on $r_1 < t \le r_2$. This follows from the relation

$$\hat{x}(t) = x_0 + \int_{t_0}^{t} f(s, \hat{x}(s))\, ds, \qquad r_1 < t \le r_2,$$

which implies that the left-hand derivative $\dot{x}(r_2 - 0)$ exists and equals $f(r_2, \hat{x}(r_2))$, which is finite.

But B is open, and by Theorem 1.3.1 there exists a solution $\varphi(t)$ of (3) passing through the point $(r_2, x(r_2 - 0))$ and defined on some interval $r_2 - \alpha \le t \le r_2 + \alpha$, for some $\alpha > 0$. Now define the function $y(t)$ as follows:

$$y(t) = \hat{x}(t), \qquad r_1 < t \le r_2,$$

$$y(t) = \varphi(t), \qquad r_2 \le t \le r_2 + \alpha,$$

and we assert that $y(t)$ is a solution for $r_1 \le t \le r_2 + \alpha$.

By uniqueness, $\hat{x}(t) = \varphi(t)$ for $r_2 - \alpha \le t \le r_2$ (we may assume that $r_1 \le r_2 - \alpha$), so we need only show the existence and continuity of $\dot{y}(t)$ at $t = r_2$. But

$$y(t) = \hat{x}(r_2) + \int_{r_2}^{t} f(s, y(s))\, ds, \qquad r_2 \le t \le r_2 + \alpha,$$

since $\varphi(t)$ is a solution, and furthermore

$$\hat{x}(r_2) = x_0 + \int_{t_0}^{r_2} f(s, y(s))\, ds,$$

since $\hat{x}(t)$ is a solution. This gives

$$y(t) = x_0 + \int_{t_0}^{t} f(s, y(s))\, ds, \qquad r_1 \le t \le r_2 + \alpha,$$

and since f and $y(t)$ are continuous this implies that $\dot{y}(t) = f(t, y(t))$, $r_1 \le t \le r_2 + \alpha$, where one-sided derivatives are intended at the end points. Therefore $y(t)$ is a continuation (to the right) of $x(t)$, and this completes the proof.

Example: Let $f(t, x) = x^2$ and let B be the set

$$B = \{(t, x) \mid -\infty < t < \infty, |x| < \infty\}.$$

For the solution $x(t) = -t^{-1}$, $0 < \alpha < t < \infty$, $\lim_{t \to \alpha + 0} x(t)$ exists and $(\alpha, x(\alpha + 0))$ belongs to B. Therefore the solution can be continued to the left, but it cannot be continued to the interval $0 \le t < \infty$, since $(0, x(0 + 0))$ is not in B.

Given the solution $x(t)$ of (3) satisfying $x(t_0) = x_0$, and whose graph lies in the domain B, let (m_1, m_2) be its maximum interval of existence (see Section 1.4). By maximality the solution cannot be extended to the right or left of (m_1, m_2). If m_2 is finite, we would expect that as t approaches m_2 from the left the values of $x(t)$ become infinite or approach the boundary of B. The following theorem indicates that this is the case.

THEOREM 6.2.2. *Let $f(t, x)$ satisfy the hypotheses of Theorem 1.3.1 in some domain B contained in R^{n+1}. Let $x = x(t)$ be a solution of (3) and let (m_1, m_2) be its maximum interval of existence. If m_2 is finite and E is any closed bounded set contained in B, then there exists an $\varepsilon > 0$ such that the point $(t, x(t))$ does not belong to E if $t > m_2 - \varepsilon$. A similar statement holds if m_1 is finite.*

Proof. Since E is closed and bounded and $R^{n+1} - B = c(B)$ is closed, an elementary topological result gives

$$\operatorname{dist}(E, c(B)) = \inf \|x - y\| = p > 0,$$

where the infimum is taken over all points x in E and all points y in $c(B)$. Therefore if (t_0, x_0) is in E, then the relation

$$\|(t_0, x_0) - (t, x)\| < p$$

implies that (t, x) is in B.

Let E^* be the closed bounded set of all points whose distance from E is less than or equal to $p/2$. Then E^* contains E and is contained in B, and there exist positive constants m and k such that

$$\|f(t, x)\| \le m, \qquad \|f(t, x_1) - f(t, x_2)\| \le nk\,\|x_1 - x_2\|$$

for all points (t, x), (t, x_1) and (t, x_2) in E^*.

Choose $a > 0$ and $b > 0$ such that $a + b < p/2$. As in the proof of Theorem 1.3.1, choose $r > 0$ satisfying the relations

$$(i)\ r \le a, \qquad (ii)\ r \le b/m, \qquad (iii)\ r < 1/nk.$$

Since the choice of m and k depended on E^*, given any point (t_0, x_0) in E, the solution $x = \varphi(t)$ of (3) satisfying $\varphi(t_0) = x_0$ will be defined for $|t - t_0| \le r$. We assert that the choice of $\varepsilon = r$ gives the desired result.

For suppose the given solution $x(t)$ has the property that $(t_1, x(t_1))$ is in E for some value of $t_1 > m_2 - r$. Then by uniqueness $x(t)$ must agree with the solution $x = \varphi(t)$, satisfying the initial conditions $\varphi(t_1) = x(t_1)$ and defined for $|t - t_1| \le r$. Since $m_2 < t_1 + r$, this contradicts the assumption that (m_1, m_2) is a maximal interval of existence for $x(t)$.

Examples

(*a*) Let B be the half-plane

$$B = \{(t, x) \mid 0 < t < \infty, x > 0\}.$$

Then the solution $x(t) = t^{1/2}$ of the equation $\dot{x} = 2/x$ escapes from every closed bounded set in B because $\lim_{t \to 0+0} x(t) = 0$. In this case $m_1 = 0$ and $m_2 = \infty$.

(*b*) Let B be the entire tx-plane. Then the solution $x(t) = \tan(t - h)$ of the equation $\dot{x} = 1 + x^2$ escapes from every closed bounded set because it becomes unbounded. In this case,

$$m_1 = -\frac{\pi}{2} + h \qquad \text{and} \qquad m_2 = \frac{\pi}{2} + h.$$

To conclude this section we will give the proof of Theorem 2.1.1, which we restate for the readers convenience.

THEOREM. *Given the equation* $\dot{x} = A(t)x + B(t)$, *where* $A(t) = (a_{ij}(t))$, $i, j = 1, \ldots, n$, *and* $B(t) = (b_1(t), \ldots, b_n(t))$ *are continuous on* $r_1 < t < r_2$, *then for any initial value* (t_0, x_0), $r_1 < t_0 < r_2$, *there exists a solution defined on* $r_1 < t < r_2$ *and satisfying the initial value.*

Proof. We proceed as in the proof of existence and uniqueness of solutions. Therefore $f(t, x) = A(t)x + B(t)$ and B is the set

$$B = \{(t, x) \mid r_1 < t < r_2,\ -\infty < x_i < \infty,\ i = 1, \ldots, n\}.$$

Given (t_0, x_0) in B and any continuous function $x = x(t)$, $r_1 < t < r_2$, the mapping $\mathscr{A}$ is then defined by

$$\begin{aligned} y = y(t) &= \mathscr{A}x \\ &= x_0 + \int_{t_0}^{t} [A(s)x(s) + B(s)]\, ds, \qquad r_1 < t < r_2. \end{aligned}$$

Since $A(t)$ and $B(t)$ are continuous on $r_1 < t < r_2$, it follows that $\mathscr{A}$ is a map from $C[r_1, r_2]$ into itself.

Let s_1 and s_2 be any real numbers such that $r_1 < s_1 < t_0 < s_2 < r_2$, and let the initial approximation be $x_0(t) = x_0$, $r_1 < t < r_2$. We will show that the successive approximations $x_{j+1}(t) = \mathscr{A}x_j$, $j = 0, 1, 2, \ldots$, converge uniformly on $s_1 \leq t \leq s_2$. Since s_1 and s_2 are arbitrary, this will imply that the solution $x(t) = \lim_{j\to\infty} x_j(t)$ is defined on $r_1 < t < r_2$, which is the desired result.

Since $A(t)$ and $x_1(t)$ are continuous on $s_1 \leq t \leq s_2$, there exist constants $k > 0$ and $c > 0$ such that

$$\|A(t)\| \leq k, \qquad \|x_1(t) - x_0\| = \|x_1(t) - x_0(t)\| \leq c,$$

for $s_1 \leq t \leq s_2$. Therefore for $s_1 \leq t \leq s_2$ we have

$$\begin{aligned} \|x_2 - x_1\| &= \left\| \int_{t_0}^{t} [A(s)x_1(s) - A(s)x_0(s)]\, ds \right\| \\ &\leq \int_{t_0}^{t} \|A(s)\|\, \|x_1(s) - x_0(s)\|\, ds \leq kc\,|t - t_0|, \end{aligned}$$

$$\begin{aligned} \|x_3 - x_2\| &\leq \int_{t_0}^{t} \|A(s)\|\, \|x_2(s) - x_1(s)\|\, ds \\ &\leq k \int_{t_0}^{t} kc\,|t - t_0|\, dt \leq \frac{k^2 c\,|t - t_0|^2}{2!}, \end{aligned}$$

and in general

$$\|x_j - x_{j+1}\| \leq c\,\frac{(k\,|t - t_0|)^j}{j!} \leq c\,\frac{[k(r_2 - r_1)]^j}{j!},$$

which implies that

$$\max_{s_1 \leq t \leq s_2} \|x_j - x_{j+1}\| = \|x_j - x_{j+1}\|_C \leq c\,\frac{[k(r_2 - r_1)]^j}{j!}.$$

But the last expression is the jth term of the series expansion for $c\exp[k(r_2 - r_1)]$, and it approaches zero as j approaches infinity. This implies that the sequence $x_j(t)$, $j = 1, 2, \ldots$, of successive approximations converges uniformly on $s_1 \leq t \leq s_2$, which completes the proof.

6.3 The Dependence of Solutions on Parameters and Approximate Solutions

We first will consider the differential equation

$$\dot{x} = f(t, x, \lambda), \tag{4}$$

where t is a scalar variable, $x = (x_1, \ldots, x_n)$, and $\lambda = (\lambda_1, \ldots, \lambda_v)$. Here $f = (f_1, \ldots, f_n)$ is an n-dimensional vector function defined on some region contained in R^{n+v+1}. The vector λ may be thought of as representing a set of parameters; therefore, given fixed $\lambda = \lambda_0$, we will denote a solution of (4) by $x = x(t, \lambda_0)$.

If f satisfied the hypotheses guaranteeing existence and uniqueness of solutions for some domain B in R^{n+v+1}, then we might expect that given a solution $x = x(t, \lambda)$ of (4) satisfying $x(t_0, \lambda) = x_0$, then varying λ slightly would only vary the solution slightly. This is equivalent to saying that solutions of (4) are continuous functions of the parameter λ. The following theorem indicates that this is certainly true locally.

THEOREM 6.3.1. *Let the equation* (4) *be given where* $f(t, x, \lambda)$ *and* $\partial f/\partial x_i$ *are defined and continuous in some domain* B *contained in* R^{n+v+1}. *If* (t_0, x_0, λ_0) *belongs to* B, *then there exist positive numbers* r *and* p *such that* (*i*) *given any* λ *such that* $\|\lambda - \lambda_0\| \leq p$, *there exists a unique solution* $x = x(t, \lambda)$ *of* (4), *defined for* $|t - t_0| \leq r$ *and satisfying* $x(t_0, \lambda) = x_0$,

(*ii*) *the solution* $x = x(t, \lambda)$ *is a continuous function of* t *and* λ.

Proof. There exist positive numbers a, b, and p such that the closed bounded set

$$P = \{(t, x, \lambda) \,|\, |t - t_0| \leq a,\ \|x - x_0\| \leq b,\ \|\lambda - \lambda_0\| \leq p\}$$

belongs to B. Furthermore, there exist positive numbers m and k such that

$$\|f(t, x, \lambda)\| \leq m, \qquad \left\|\frac{\partial f_i(t, x, \lambda)}{\partial x_j}\right\| \leq k,$$

$i, j = 1, \ldots, n$, whenever (t, x, λ) belongs to P.

As before, choose $r > 0$ such that

(*i*) $r \leq a$, (*ii*) $r \leq b/m$, (*iii*) $r < 1/nk$,

and let P_r be the set

$$P_r = \{(t, x, \lambda) \,|\, |t - t_0| \leq r,\ \|x - x_0\| \leq b,\ \|\lambda - \lambda_0\| \leq p\}.$$

Let $\mathscr{C}_r$ be the set of all continuous functions $x = x(t, \lambda)$ having the property that the point $(t, x(t, \lambda))$ belongs to P_r for $|t - t_0| \leq r$ and $\|\lambda - \lambda_0\| \leq p$. This is equivalent to the relation $\|x(t, \lambda) - x_0\| \leq b$ for $|t - t_0| \leq r$ and $\|\lambda - \lambda_0\| \leq p$.

Then for any function $x = x(t, \lambda)$ in $\mathscr{C}_r$ let the map $\mathscr{A}_\lambda$ be defined by

$$\mathscr{A}_\lambda x = x_0 + \int_{t_0}^{t} f(s, x(s, \lambda))\, ds, \qquad |t - t_0| \leq r.$$

If we fix λ and let the first approximation be $x_0(t, \lambda) = x_0$, then the choice of r guarantees that the sequence of successive approximations $x_{j+1} = \mathscr{A}_\lambda x_j$, $j = 0, 1, 2, \ldots$, converges uniformly on $|t - t_0| \leq r$ to $x = x(t, \lambda)$, a unique solution of (4) satisfying $x(t_0, \lambda) = x_0$.

Furthermore, for $s > m$ we have

$$\|x_m - x_s\|_C \leq \frac{\alpha^m}{1 - \alpha} \|x_0 - x_1\|_C \leq \frac{\alpha^m b}{1 - \alpha}, \qquad 0 < \alpha < 1,$$

for $|t - t_0| \le r$ and $\|\lambda - \lambda_0\| \le p$. This shows that the sequence of successive approximations converges uniformly in λ as well, and hence $x = x(t, \lambda)$ is a continuous function of λ as well.

To conclude this chapter we wish to discuss briefly two results related to the dependence of solutions on parameters or on initial conditions. The method of proof is merely indicated.

First of all suppose we consider the equation (4), where t and λ are scalar quantities and $x = (x_1, \ldots, x_n)$ is a vector variable. Given (t_0, x_0, λ_0), suppose that we know the solution $x = x(t, \lambda_0)$ satisfying $x(t_0, \lambda_0) = x_0$, and for λ near λ_0 we wish to find an approximation to the solution $x = x(t, \lambda)$ satisfying $x(t_0, \lambda) = x_0$.

For $|\lambda - \lambda_0|$ sufficiently small we may use the approximation

$$x(t, \lambda) = x(t, \lambda_0) + (\lambda - \lambda_0)y(t), \tag{5}$$

where

$$y(t) = \left.\frac{\partial x(t, \lambda)}{\partial \lambda}\right|_{\lambda = \lambda_0}$$

This derivative will exist, for instance, if f is continuously differentiable with respect to the variable λ. Substitute the expression (5) for $x(t, \lambda)$ in (4) and differentiate with respect to λ and set $\lambda = \lambda_0$. This shows that $y(t)$ satisfies the relation

$$\dot{y} = A(t)y + B(t),$$

where

$$A(t) = \left(\frac{\partial f_i}{\partial x_j}\right)_{\substack{\lambda = \lambda_0 \\ x = x(t, \lambda_0)}} \qquad B(t) = \left(\frac{\partial f_1}{\partial \lambda}, \ldots, \frac{\partial f_n}{\partial \lambda}\right)_{\substack{\lambda = \lambda_0 \\ x = x(t, \lambda_0)}}$$

This is the *variational equation* for the differential equation (4), and evidently is a first-order nonhomogeneous system with initial conditions $y(t_0) = 0$, since $x(t_0, \lambda) = x(t_0, \lambda_0)$. If we can solve or obtain an approximation to $y(t)$ in some neighborhood of t_0, then the relation (5) will give us an approximation to the solution $x = x(t, \lambda)$.

Example: Let $n = 1$ and consider the equation $\dot{x} = \lambda x^2 + t$. Letting $\lambda_0 = t_0 = x_0 = 0$, we have

$$x(t, \lambda_0) = x(t, 0) = t^2/2, \qquad x(0, 0) = 0.$$

Furthermore, $\partial f/\partial x = 2x\lambda$ and $\partial f/\partial \lambda = x^2$ and the variational equation then becomes

$$\dot{y} = t^4/4, \qquad y(0) = 0,$$

whose solution is $y(t) = t^5/20$. Then for λ near zero we have the approximation

$$x(t, \lambda) = \frac{1}{2} t^2 + \frac{\lambda}{20} t^5.$$

Finally, suppose we are given two differential equations with initial conditions

$$\dot{x} = f(t, x), \qquad \dot{y} = g(t, y),$$
$$x(t_0) = x_0, \qquad y(t_0) = y_1,$$

where we assume that f and g are defined in some domain B in R^{n+1}, and

$$\sup \|f(t, x) - g(t, x)\| < \varepsilon, \qquad (t, x) \text{ in } B,$$

where ε is sufficiently small. Suppose that we know the solution of the second equation and that $\|x_0 - y_1\| < \delta$ with δ sufficiently small. We wish to obtain an estimate of the error obtained by replacing the first equation by the second.

We will assume that $f(t, x)$ is continuous and satisfies the Lipschitz condition

$$\|f(t, x_1) - f(t, x_2)\| \leq k \|x_1 - x_2\|$$

for (t, x_1), (t, x_2) in B. Our assumptions lead to the following estimates:

$$\begin{aligned}
\|x(t) - y(t)\| &\leq \|x_0 - y_1\| + \left\| \int_{t_0}^{t} \Big[f(s, x(s)) - g(s, y(s)) \Big] \, ds \right\| \\
&\leq \delta + \int_{t_0}^{t} \|f(s, y(s)) - g(s, y(s))\| \, ds \\
&\quad + \int_{t_0}^{t} \|f(s, x(s)) - f(s, y(s))\| \, ds \\
&\leq \delta + \varepsilon |t - t_0| + k \int_{t_0}^{t} \|x(s) - y(s)\| \, ds.
\end{aligned}$$

If we let

$$M = \sup \|x(t) - y(t)\|, \qquad |t - t_0| \le r,$$

with r sufficiently small, we have

$$\|x(t) - y(t)\| \le \delta + \varepsilon|t - t_0| + kM|t - t_0|$$

for $|t - t_0| \le r$.

Now substitute this expression on the right side of the above inequality. This gives the relation

$$\|x(t) - y(t)\| \le \delta + k\delta|t - t_0| + \varepsilon|t - t_0| + \varepsilon\frac{k|t - t_0|^2}{2!} + M\frac{k^2|t - t_0|^2}{2!}.$$

Repeating the substitution n times, we have the relation

$$\|x(t) - y(t)\| \le \delta\sum_{j=0}^{n-1}\frac{k^j|t - t_0|^j}{j!} + \varepsilon\sum_{j=1}^{n}\frac{k^{j-1}|t - t_0|^j}{j!} + M\frac{k^n|t - t_0|^n}{n!}$$

for $|t - t_0| \le r$. The last term on the right goes to zero uniformly as n approaches ∞, and this leads to the estimate

$$\|x(t) - y(t)\| \le \delta\exp(k|t - t_0|) + \frac{\varepsilon}{k}[\exp(k|t - t_0|) - 1].$$

Example: We consider the pair of differential equations

$$\dot{x} = f(t, x) = 1 + x^2 + t^2,$$

$$\dot{y} = g(t, y) = 1 + y^2,$$

with the same initial conditions $x(0) = y(0) = 0$. Hence $\delta = 0$. If we let B be the domain

$$B = \{(t, x) \mid |t| < \tfrac{1}{4}, |x| < 1\},$$

then

$$|f(t, x) - g(t, x)| = |t^2| < \tfrac{1}{16} = \varepsilon,$$

and

$$|f(t, x) - f(t, y)| = |x^2 - y^2| < 2|x - y|.$$

Hence $k = 2$. The solution of the second equation is $y(t) = \tan t$, and an estimate of the error in replacing the first equation by the second is given by

$$|x(t) - \tan t| \leq \tfrac{1}{32}[e^{1/2} - 1] = 0.0203,$$

valid for $|t| \leq r < \frac{1}{4}$.

The estimate above can also be used to measure the change in a solution due to a change in the initial conditions: in this case $\varepsilon = 0$.

Problems

1. Use the method of successive approximations to find the first three approximations to solutions of the following equations.

(a) $\dot{x} = t^2 - x$, $x(0) = 1$.

(b) $\dot{x} = t^2 - x$, $x(1) = 2$.

(c) $\dot{x} = 1 + t^2 + x^2$, $x(0) = 0$.

(d) $\dot{x} = x^2 - t$, $\quad x(1) = 1$.

(e) $\dot{x} = t + y$, $x(0) = 2$.
$\dot{y} = t - x^2$, $y(0) = 1$.

2. (a) Using the inequality

$$\|x - x_j\| \leq \sum_{s=j}^{\infty} \|x_{s+1} - x_s\|,$$

show that an upper bound for the error in stopping at the jth successive approximation to a solution is given by

$$\|x - x_j\|_C \leq \frac{m}{nk} e^{nkr} \frac{(nkr)^{j+1}}{(j+1)!},$$

where m, n, k and r are as in the proof of Theorem 1.3.1.

(b) For $|t| < \frac{1}{2}$ and $|x| < \frac{1}{2}$, show that the error in stopping at the third approximation of equation 1(c) is less than 1.08×10^{-3}.

3. Given the scalar equation

$$\dot{x} = f(t, x), \qquad x(t_0) = x_0,$$

and a positive integer N, some simple techniques for obtaining in N steps an approximate value x_N of $x(t_0 + T)$, $T > 0$, are as follows.

(i) Euler-Cauchy method

$$x_{n+1} = x_n + hf(t_n, x_n), \qquad n = 0, 1, \ldots, N-1.$$

(*ii*) Taylor series method of order 2:

$$x_{n+1} = x_n + hf(t_n, x_n) + \frac{h^2}{2}[f_t(t_n, x_n) + f_x(t_n, x_n)f(t_n, x_n)],$$
$$n = 0, 1, \ldots, N-1.$$

(*iii*) Modified Euler method:

$$x_{n+2} = x_n + 2hf(t_{n+1}, x_{n+1}), \qquad n = 0, 1, \ldots, N-2,$$

where x_1 is obtained by some other method. The number x_n is the approximate value of $x(t_n)$, where $t_{n+1} - t_n = h = T/N$.
(a) Use the methods (*i*) and (*iii*) with x_1 obtained using (*ii*) and $N = 10$ to approximate the value $x(1)$ of the solution of $\dot{x} = x$, $x(0) = 1$.
(b) Use the methods (*i*) and (*ii*) with $N = 10$ and $N = 5$ respectively to approximate the value $x(1)$ of the solution of $\dot{x} = t^2 + x$, $x(0) = 1$.
(c) Use the methods (*i*) and (*iii*) with x_1 obtained using (*ii*) and $N = 5$ to approximate the value $x(1.50)$ of the solution of $\dot{x} = 1 + x^2$, $x(1.45) = 8.238$.
In all cases compare the results with the actual values of the solution.

4. Let $f(t, x)$ satisfy the hypotheses of Theorem 1.3.1 in the strip $-\infty < t < \infty$, $a \leq x \leq b$, and suppose that f is periodic in t with period T. If

$$f(t, a) > 0, \qquad f(t, b) < 0, \qquad -\infty < t < \infty,$$

show that this implies that the equation $\dot{x} = f(t, x)$ has a periodic solution of period T. (*Hint*: First show that any solution $x(t)$ satisfying $x(0) = x_0$, $a \leq x_0 \leq b$, is defined for $0 \leq t \leq T$ and takes on values between a and b. Then show there exists such a solution satisfying $x(0) = x(T)$ and that its periodic continuation is the required solution.)

5. Show that if $f(t, x)$ has continuous partial derivatives up to order m, then any solution of $\dot{x} = f(t, x)$ has continuous derivatives up to order $m + 1$.

6. Given the equation

$$\dot{x} = x + \lambda(t + x^2), \qquad |\lambda| < 0.1,$$

and let the solution $x(t, \lambda_0) = x(t, 0)$ satisfy the initial condition $x(0, 0) = 1$. Find the solution of the variational equation, and use it to obtain an esimate for $0 \leq t < \frac{1}{2}$ of the difference between the solutions $x(t, 0)$ and $x(t, \lambda)$, where $x(0, \lambda) = 1$.

7. Given the equation

$$\dot{x} = t^2 + e^t \sin x, \qquad x(0) = 0 = x_0,$$

estimate the variation of the solution for $0 \leq t < \frac{1}{4}$ if x_0 is perturbed by 0.01.

8. Find the error in using the approximate solution

$$x(t) = \exp\left(-\frac{t^3}{6}\right)$$

for the equation

$$\ddot{x} + tx = 0, \qquad x(0) = 1, \qquad \dot{x}(0) = 0,$$

where $|t| \leq \frac{1}{2}$.

APPENDIX

A

Series Solutions of Second-Order Linear Equations

In general, given the nth order homogenous equation

$$y^{(n)} + a_1(t)y^{(n-1)} + \cdots + a_n(t)y = 0, \tag{1}$$

there is no method of finding a set of fundamental solutions, and hence a solution, explicitly. The following approach is often useful. We assume that a solution of (1) can be expressed in the form

$$y(t) = (t - t_0)^r \sum_{j=0}^{\infty} \alpha_j(t - t_0)^j, \qquad \alpha_j \text{ constant}, \tag{2}$$

where the power series in the expression is convergent in some neighborhood $|t - t_0| < a$ of t_0. In the case $r < 0$, then, the expression will be valid in the deleted neighborhood $0 < |t - t_0| < a$.

Differentiating the expression by terms, and substituting it in (1) will hopefully give us a solvable recurrence relation for each α_j in terms of its predecessors, and specifically in terms of the first n α_j's; hence $\alpha_j = \varphi_j(\alpha_0, \ldots, \alpha_{n-1})$, $j \geq n$. The expression (2) is then a formal solution, and we must verify the convergence of the series to assure us that it is in fact a solution. The arbitrary constants $\alpha_0, \ldots, \alpha_{n-1}$ will be prescribed by given initial values of the solution.

The described method is especially fruitful in the case $n = 2$—that is, for second-order linear equations—and we proceed to discuss it in detail. For the sake of generality, we will assume that all functions concerned are complex-valued functions of the complex variable z. Our equation then becomes

$$\frac{d^2\omega}{dz^2} + p(z)\frac{d\omega}{dz} + q(z)\omega = 0, \tag{3}$$

and we will attempt to find a solution $\omega(z)$ of the form

$$\omega(z) = (z - z_0)^r \sum_{n=0}^{\infty} a_n(z - z_0)^n, \tag{4}$$

where the power series converges in some neighborhood of the point z_0. We require some preliminary definitions.

DEFINITION: A function $\varphi(z)$ is said to be analytic at $z = z_0$ if it has a Taylor's series expansion $\varphi(z) = \sum_{n=0}^{\infty} b_n(z - z_0)^n$, which converges in some neighborhood of $z = z_0$.

DEFINITION: A function $\theta(z)$ is said to possess a pole of order k at $z = z_0$ if $\theta(z) = \varphi(z)(z - z_0)^{-k}$, where $\varphi(z)$ is analytic at $z = z_0$.

If either of the functions $p(z)$, $q(z)$ in (3) is not analytic at $z = z_0$, then we say that $z = z_0$ is a *singular point* of (3). The class of singular points that we will consider is defined as follows.

DEFINITION: The singular point $z = z_0$ is said to be a *regular singular point* of (3) if $p(z)$ and $q(z)$ possess at most a pole of order 1 and 2 respectively at $z = z_0$.

DEFINITION: The point $z = \infty$ is said to be a regular singular point of (3) if, under the transformation $z = 1/\xi$, the point $\xi = 0$ is a regular singular point of the transformed equation.

Remark: We may easily verify that the transformed equation is given by

$$\frac{d^2\omega}{d\xi^2} + \left[\frac{2}{\xi} - \frac{1}{\xi^2}\, p\left(\frac{1}{\xi}\right)\right]\frac{d\omega}{d\xi} + \frac{1}{\xi^4}\, q\left(\frac{1}{\xi}\right)\omega = 0.$$

From the above definitions it follows that if z_0 is a regular singular point of (3), we may write the equation in the form

$$\frac{d^2\omega}{dz^2} + \frac{P(z)}{z - z_0}\frac{d\omega}{dz} + \frac{Q(z)}{(z - z_0)^2}\,\omega = 0, \tag{3'}$$

where $P(z)$ and $Q(z)$ are analytic at $z = z_0$. Therefore we may write

$$P(z) = \sum_{n=0}^{\infty} p_n(z - z_0)^n,\ Q(z) = \sum_{n=0}^{\infty} q_n(z - z_0)^n,$$

where both series converge in $|z - z_0| < a,\ a > 0$. To simplify computations we assume that $a_0 = 1$ in (4).

THEOREM A.1. *If z_0 is a regular singular point of* (3), *then there exists a solution of the form*

$$\omega(z) = (z - z_0)^r\left[1 + \sum_{n=1}^{\infty} a_n(z - z_0)^n\right], \tag{5}$$

and the above expression is valid in $0 < |z - z_0| < a$.

Proof. Recalling that

$$\left(\sum_0^\infty a_n z^n\right)\left(\sum_0^\infty b_n z^n\right) = \sum_{n=0}^\infty\left(\sum_{k=0}^n a_k b_{n-k}\right)z^n,$$

we substitute the expression (5) for $\omega(z)$ in (3′), which gives us

$$(z - z_0)^r\left\{r^2 + r(p_0 - 1) + q_0 + \sum_{n=1}^{\infty}\left[a_n(r + n)(r + n - 1) + rp_n\right.\right.$$
$$\left.\left. + \sum_{k=0}^{n-1} a_{n-k}(r + n - k)p_k + q_n + \sum_{k=0}^{n-1} a_{n-k}\,q_k\right](z - z_0)^n\right\} = 0.$$

Let $F(r + n) = (r + n)^2 + (p_0 - 1)(r + n) + q_0$, $n = 0, 1, 2, \ldots,$ and set $k = 0$, and the above expression can be written

$$(z - z_0)^r\left\{F(r) + \sum_{n=1}^{\infty}\left[a_n F(r + n) + \psi_n\right](z - z_0)^n\right\} = 0, \tag{6}$$

where ψ_n is a term depending on $a_1, \ldots, a_{n-1}, p_k$, and q_k, $k = 1, \ldots, n$.

Each coefficient must be equal to zero, and hence

$$F(r) = r^2 + (p_0 - 1)r + q_0 = 0. \tag{7}$$

This is called the *indicial equation* of (3′) at $z = z_0$. Let r_1 and r_2 be the two roots of $F(r)$ and let $s = r_1 - r_2$ be such that $\text{Re}(s) \geq 0$. Since $F(r_1) = 0$ and $p_0 - 1 = -(r_1 + r_2) = -2r_1 + s$, we have for all n

$$\begin{aligned} F(r_1 + n) &= F(r_1) + 2nr_1 + n^2 + (p_0 - 1)n \\ &= n^2 + n(p_0 - 1 + 2r_1) = n(s + n) \neq 0. \end{aligned}$$

Therefore we can solve the recurrence relation $a_n F(r_1 + n) + \psi_n = 0$ for a_n in terms of $a_1, \ldots, a_{n-1}, p_k$, and q_k, $k = 1, \ldots, n$; hence (5) with $r = r_1$ represents a formal solution of (3′), and hence of (3).

To prove that it represents an actual solution, we need only show that the series $\sum_{n=1}^{\infty} a_n(z - z_0)^n$ has a positive radius of convergence. Then, since the only singularity of $p(z)$ and $q(z)$ in $|z - z_0| < a$ occurs at $z = z_0$, by an analogue of Theorem 2.2.1 the solution $\omega(z)$ is defined in $0 < |z - z_0| < a$, and the expression (5) for $\omega(z)$ is valid there.

Choose α such that $0 < \alpha < a$, and we can find $K > 1$ such that $|p_n|$, $|q_n|$, and $|r_1 p_n + q_n|$ are less than K/α^n for all n. Since $|n + s| \geq n$ for any n, we have

$$|a_1| = \left|\frac{\psi_1}{F(r_1 + 1)}\right| = \left|\frac{r_1 p_1 + q_1}{s + 1}\right| \leq \frac{K}{\alpha}.$$

Suppose that $|a_n| \leq K^n/\alpha^n$ for $n = 1, \ldots, m - 1$; then by computing ψ_m we have

$$\begin{aligned} |a_m| &= \left|\frac{\psi_m}{F(r_1 + m)}\right| \\ &\leq \frac{\left[\sum_{k=1}^{m-1} |a_{m-k}|\,|r_1 p_k + q_k| + |r_1 p_m + q_m| + \sum_{k=1}^{m-1} (m - k)\,|a_{m-1}|\,|p_k|\right]}{m\,|s + m|} \\ &\leq \left[\frac{m + \dfrac{m(m-1)}{2}}{m^2}\right] \frac{K^m}{\alpha^m} \leq \frac{K^m}{\alpha^m}. \end{aligned}$$

Hence $|a_n| \leq K^n/\alpha^n$ for all n, which implies that $\sum_{n=1}^{\infty} a_n(z - z_0)^n$ has a radius of convergence at least as large as $\alpha/K > 0$. This proves the theorem.

We now discuss the possibility of obtaining another solution, corresponding to the root r_2 of the indicial equation. In addition, the two solutions constructed will be linearly independent, and hence all solutions of (3) can be obtained.

COROLLARY. *If z_0 is a regular singular point of (3), then corresponding to the root r_2 of the indicial equation there exists another solution of (3) of the form*

$$\omega(z) = (z - z_0)^{r_2}\left[1 + \sum_{n=1}^{\infty} b_n(z - z_0)^n\right],$$

or of the form

$$\omega(z) = \omega_1(z)\beta \log(z - z_0) + (z - z_0)^{r_2} \sum_{n=0}^{\infty} b_n(z - z_0)^n,$$

where $\omega_1(z)$ is the solution corresponding to the root r_1 and β is a constant. The expressions are valid in $0 < |z - z_0| < a$.

Proof. Two cases occur depending on the value of $s = r_1 - r_2$.

Case 1. $s \neq 0$, positive integer. In this case $F(r_2 + n) = n(-s + n) \neq 0$, and we can solve the recurrence relation $b_n F(r_2 + n) + \Omega_n = 0$, where Ω_n corresponds to the previous ψ_n. If $\lambda = \sup_{n \geq 1} |1 - s/n|^{-1}$, it can be shown that for $0 < \alpha < a$ we have $|b_n| \leq (M\lambda)^n/\alpha^n$ for all n, with M defined as before, and the result follows.

Case 2. $s = 0$, or $s =$ positive integer. If $s = 0$, then $r_1 = r_2$, so nothing new is gained; if $s = m$, then $F(r_2 + m) = 0$ and the recurrence relation cannot be solved. We proceed by the method of reduction of order and assume there exists a solution of the form $\omega_2(z) = \omega_1(z) \int^z \varphi(u)\, du$, where $\omega_1(z)$ is the solution corresponding to $r = r_1$.

Substitution in (3′) leads to the equation

$$\frac{d\varphi}{dz} + \left(\frac{P(z)}{z - z_0} + \frac{2\dfrac{d\omega_1}{dz}}{\omega_1}\right)\varphi = 0.$$

The term in parentheses can be expressed in the form

$$\frac{-(p_0 + 2r_1)}{z - z_0} + \varphi_2(z),$$

where $\varphi_2(z)$ is analytic at z_0, and hence

$$\begin{aligned}\varphi(z) &= \exp\left\{\int^z \left[\frac{-(p_0 + 2r_1)}{u - z_0} + \varphi_2(u)\right] du\right\}\\ &= (z - z_0)^{-p_0 - 2r_1}\varphi_3(z),\end{aligned}$$

where $\varphi_3(z)$ is analytic at z_0. Since $-p_0 - 2r_1 = -s - 1$,

$$\varphi(z) = (z - z_0)^{-s-1} \sum_{n=0}^{\infty} \beta_n (z - z_0)^n,$$

and now integrating $\varphi(z)$ we have (if $\beta = \beta_s \neq 0$)

$$\int^z \varphi(u)\, du = \beta \log(z - z_0) + (z - z_0)^{-s}\varphi_4(z),$$

where $\varphi_4(z)$ is analytic at $z = z_0$.

We write

$$\sum_{n=0}^{\infty} b_n (z - z_0)^n = \varphi_4(z)\left\{1 + \sum_{n=1}^{\infty} a_n (z - z_0)^n\right\},$$

and since $-s + r_1 = r_2$, we finally have

$$\omega_2(z) = \omega_1(z)\beta \log(z - z_0) + (z - z_0)^{r_2} \sum_{n=0}^{\infty} b_n (z - z_0)^n,$$

which is the required result. If $\beta = \beta_s = 0$, then $\omega_2(z)$ is of the form given in Case 1. Note that if $s = 0$, the logarithmic form always appears.

In summary, to find a solution of (3) that is valid near a regular point z_0, we solve the indicial equation to find r_1, then substitute in (3) a series

$$(z - z_0)^{r_1}\left\{1 + \sum_{n=1}^{\infty} a_n (z - z_0)^n\right\}$$

The recurrence relation is solvable and we can find the a_n explicitly. To find a second linearly independent solution, we must consider the value of s. The examples below illustrate several cases.

We now briefly discuss the case in which the coefficients $p(z)$ and $q(z)$ in the equation (3) are analytic at $z = z_0$.

DEFINITION: The point $z = z_0$ is said to be an *ordinary point* of (3) if $p(z)$ and $q(z)$ are analytic at $z = z_0$.

It follows then that $p(z)$ and $q(z)$ are expressible in power series

$$p(z) = \sum_{n=0}^{\infty} b_n(z - z_0)^n, \qquad q(z) = \sum_{n=0}^{\infty} c_n(z - z_0)^n,$$

which converge in some neighborhood of $z = z_0$, say for $|z - z_0| < a$. If we substitute the series expression for $p(z)$ and $q(z)$ in (3) and assume a solution of the form

$$\omega(z) = \sum_{n=0}^{\infty} a_n(z - z_0)^n,$$

we are led to the recurrence relation

$$(n + 2)(n + 1)a_{n+2} = -\sum_{k=0}^{n} [(k + 1)b_{n-k}\, a_{k+1} + c_{n-k}\, a_k].$$

Letting a_0 and a_1 be arbitrary, we can solve for a_n, $n \geq 2$, in terms of them.

It can be shown that the resulting power series for $\omega(z)$ converges for $|z - z_0| < a$, and therefore represents a solution. In most cases, since a_0 and a_1 are arbitrary, we can obtain two linearly independent solutions. In any case, since $a_0 = \omega(z_0)$ and $a_1 = \omega'(z_0)$, we can determine the solution satisfying any given initial conditions.

Examples

(*a*) The equation

$$2z \frac{d^2\omega}{dz^2} + \frac{d\omega}{dz} - \omega = 0$$

can be written in the form

$$\frac{d^2\omega}{dz^2} + \frac{1/2}{z}\frac{d\omega}{dz} + \frac{-z/2}{z^2}\,\omega = 0.$$

Therefore $z = 0$ is a regular singular point, $p_0 = \frac{1}{2}$, $q_0 = 0$, and the indicial equation is $r^2 - \frac{1}{2}r = 0$. Its roots are $r_1 = \frac{1}{2}$ $r_2 = 0$; hence $s = \frac{1}{2}$ and $s \neq 0$, positive integer.

Substituting the series

$$\omega_1(z) = \sum_0^\infty a_n z^{n+(1/2)}$$

in the original equation gives

$$\sum_0^\infty 2(n + \tfrac{1}{2})(n - \tfrac{1}{2})a_n z^{n-(1/2)} + \sum_0^\infty (n + \tfrac{1}{2})a_n z^{n-(1/2)} - \sum_0^\infty a_n z^{n+(1/2)} = 0,$$

and combining the first two series and shifting indices in the last gives

$$\sum_1^\infty [n(2n + 1)a_n - a_{n-1}]z^{n-(1/2)} = 0.$$

Therefore

$$a_n = \frac{2a_{n-1}}{(2n + 1)(2n)} \qquad \text{or} \qquad a_n = \frac{2^n a_0}{(2n + 1)!}, \qquad n \geq 1,$$

and therefore

$$\omega_1(z) = a_0 z^{1/2} \sum_0^\infty \frac{2^n}{(2n + 1)!} z^n.$$

Proceeding in a similar manner for $r_2 = 0$ gives a second linearly independent solution,

$$\omega_2(z) = b_0 \sum_0^\infty \frac{2^n}{(2n)!} z^n,$$

where a_0 and b_0 are arbitrary constants.

(*b*) The equation

$$z\frac{d^2\omega}{dz^2}+(3+z^3)\frac{d\omega}{dz}+3z^2\omega=0$$

can be written in the form

$$\frac{d^2\omega}{dz^2}+\frac{3+z^3}{z}\frac{d\omega}{dz}+\frac{3z^3}{z^2}\,\omega=0.$$

Therefore $z=0$ is a regular singular point, $p_0=3$, $q_0=0$, and the indicial equation is $r^2+2r=0$. Its roots are $r_1=0$, $r_2=-2$; hence $s=2$, a positive integer.

In this case it is advantageous to use $r=r_2$, so let

$$\omega(z)=\sum_0^\infty a_n z^{n-2}.$$

Substituting in the original equation and shifting indices gives

$$-a_1 z+\sum_3^\infty[(n-2)na_n+(n-2)a_{n-3}]z^{n-3}=0.$$

Therefore $a_1=0$ and $a_n=(-a_{n-3})/n$, which implies that

$$a_1=a_4=\cdots=a_{3n+1}=\cdots=0,$$

and

$$a_{3n}=\frac{(-1)^n a_0}{(3n)(3n-3)\cdots(3)}=\frac{(-1)^n a_0}{3^n n!},$$

$$a_{3n+2}=\frac{(-1)^n a_2}{(3n+2)\cdots 8\cdot 5},$$

where a_0 and a_0 are arbitrary. Then we have

$$\omega(z)=a_0\left[z^{-2}+\sum_1^\infty\frac{(-1)^n}{3^n n!}z^{3n-2}\right]+a_2\left[1+\sum_1^\infty\frac{(-1)^n}{(3n+2)\cdots 8\cdot 5}z^{3n}\right]$$

and the bracketed series represent a pair of linearly independent solutions corresponding to the roots of the indicial equation. Evidently this is the nonlogarithmic case.

(c) The equation

$$(z-1)\frac{d^2\omega}{dz^2}+z\frac{d\omega}{dz}+\omega=0$$

can be written in the form

$$\frac{d^2\omega}{dz^2}+\frac{1+(z-1)}{z-1}\frac{d\omega}{dz}+\frac{z-1}{(z-1)^2}\,\omega=0.$$

Therefore $z=1$ is a regular singular point, $p_0=1$, $q_0=0$, and the indicial equation is $r^2=0$; hence $s=0$, a logarithmic case.

For this case it is convenient to assume a solution of the form

$$\omega(z)=(z-1)^r+\sum_1^\infty a_n(z-1)^{n+r},$$

where r is indeterminate. Writing the equation in the form

$$(z-1)\frac{d^2\omega}{dz^2}+[1+(z-1)]\frac{d\omega}{dz}+\omega=0,$$

and substituting $\omega(z)$ as given above leads to the relation

$$r^2(z-1)^{r-1}+[(r+1)+(r+1)^2a_1](z-1)^r$$
$$+\sum_2^\infty[(n+r)^2a_n+(n+r)a_{n-1}](z-1)^{n+r-1}=0.$$

Therefore

$$a_1=\frac{-1}{r+1},\qquad a_n=\frac{-a_{n-1}}{n+r}=\frac{(-1)^n}{(n+r)(n-1+r)\cdots(1+r)},$$

and

$$\frac{\partial a_n}{\partial r}=(-1)^{n+1}\frac{\left[\dfrac{1}{n+r}+\dfrac{1}{n-1+r}+\cdots+\dfrac{1}{1+r}\right]}{(n+r)(n-1+r)\cdots(1+r)}.$$

It follows that

$$a_n\bigg|_{r=0} = \frac{(-1)^n}{n!},$$

$$\frac{\partial a_n}{\partial r}\bigg|_{r=0} = \frac{(-1)^{n+1}H_n}{n!},$$

where $H_n = \sum_1^n 1/k$. The two linearly independent solutions of the equations are then given by

$$\omega_1(z) = 1 + \sum_1^\infty \frac{(-1)^n}{n!}(z-1)^n,$$

$$\omega_2 = \omega_1(z)\log(z-1) + \sum_1^\infty \frac{(-1)^{n+1}H_n}{n!}(z-1)^n.$$

For the case $s =$ positive integer, logarithmic case, we should assume a solution of the form

$$\omega(z, r) = (z - z_0)^r(r - r_2) + \sum_1^\infty a_n(z - z_0)^{n+r},$$

then substitute and evaluate $a_n = a_n(r)$. The two linearly independent solutions are given by

$$\omega_1(z) = \omega(z, r_2), \qquad \omega_2(z) = \frac{\partial\omega(z, r)}{\partial r}\bigg|_{r=r_2}$$

Finally, it should be noted that solutions need not have singularities at a regular singular point. For instance, the point $z = z_0$ is a regular singular point of the equation

$$\frac{d^2\omega}{dz^2} + \frac{-2}{z - z_0}\frac{d\omega}{dz} + \frac{2}{(z - z_0)^2}\,\omega = 0,$$

but a fundamental pair of solutions is given by

$$\omega_1(z) = (z - z_0), \qquad \omega_2(z) = (z - z_0)^2,$$

which are analytic at $z = z_0$.

APPENDIX

B

Linear Systems with Periodic Coefficients

In this appendix we will discuss the linear system

$$\dot{x} = A(t)x, \tag{1}$$

where $x = (x_1, \ldots, x_n)$ and $A(t) = (a_{ij}(t))$ is a continuous periodic matrix defined on $-\infty < t < \infty$. Therefore there exists a nonzero number T such that

$$A(t + T) = A(t), \qquad -\infty < t < \infty.$$

It does not necessarily follow that there exists a nontrivial solution $x(t)$ of (1) having period T.

Example: For the equation

$$\dot{x} = (\cos^2 t)x,$$

$x = x(t)$ a scalar function, we have

$$a(t) = \cos^2 t, \qquad a(t + 2\pi) = a(t).$$

Any nontrivial solution is of the form

$$x(t) = x_0 \exp\left[\frac{t}{2} + \frac{\sin 2t}{4}\right]$$

and is not periodic.

A basic result (Floquet's theorem) describing the properties of solutions of (1) is the following.

THEOREM B.1. *Given the system* (1) *with* $A(t)$ *a continuous periodic matrix with period* T, *then there exists a nonzero constant* λ *(real or complex) and at least one nontrivial solution* $x(t)$ *of* (1) *having the property that*

$$x(t+T) = \lambda x(t), \qquad -\infty < t < \infty.$$

Proof. Let

$$\varphi_j(t) = (\varphi_{1j}(t), \ldots, \varphi_{nj}(t)), \qquad j = 1, \ldots, n,$$

be a fundamental system of solutions of (1). The corresponding fundamental matrix $\Phi(t) = (\varphi_{ij}(t))$ then satisfies the relation $\det \Phi(0) \neq 0$, and

$$\det \Phi(t) = \det \Phi(0) \exp\left[\int_0^t \operatorname{tr} A(s)\, ds\right], \tag{2}$$

by the results of Chapter 2. It follows that $\det \Phi(t+T) \neq 0$ and hence $\Phi(t+T)$ is also a fundamental matrix, and its columns therefore form a fundamental system of solutions.

Hence there exist constants c_{jk}, $j, k = 1, \ldots, n$ such that

$$\varphi_{ik}(t+T) = \sum_{j=1}^{n} c_{jk}\varphi_{ij}(t), \qquad i, k = 1, \ldots, n,$$

or equivalently there exists an $n \times n$ constant matrix C such that

$$\Phi(t+T) = \Phi(t)C, \qquad -\infty < t < \infty.$$

If we let $t = 0$, then (2) implies that

$$\det C = \exp\left[\int_0^T \operatorname{tr} A(s)\, ds\right] \neq 0.$$

Let λ be a characteristic root (real or complex) of the matrix C—that is, a root of the characteristic polynomial $\det(C - \lambda I)$. Then $\det C \neq 0$ implies $\lambda \neq 0$ and furthermore there exists a nonzero vector α such that $C\alpha = \lambda\alpha$.

Consider the solution $x(t) = \Phi(t)\alpha$ of (1). Then for all t we have

$$x(t + T) = \Phi(t + T)\alpha = \Phi(t)C\alpha$$
$$= \Phi(t)\lambda\alpha = \lambda\Phi(t)\alpha = \lambda x(t),$$

and conversely for any solution having this property, λ must be a characteristic root of C. If $\lambda_1, \ldots, \lambda_m$, $1 \leq m \leq n$, are the distinct characteristic roots of C, then there are at least m solutions of (1) satisfying the conclusion of the theorem. This completes the proof.

It should be noted that if $\Omega(t)$ is any other fundamental matrix, then there exists a constant matrix A, with $\det A \neq 0$, such that $\Omega(t) = \Phi(t)A$. Since $\Phi(t + T) = \Phi(t)C$, we have

$$\Omega(t + T) = \Phi(t + T)A = \Phi(t)CA = \Omega(t)A^{-1}CA.$$

From linear algebra, we know that the characteristic roots of C and $A^{-1}CA$ are the same, and therefore the numbers $\lambda_1, \ldots, \lambda_m$, $1 \leq m \leq n$, are independent of the choice of a fundamental matrix.

DEFINITION: The distinct characteristic roots $\lambda_1, \ldots, \lambda_m$ of the matrix C are called the *characteristic factors* or *multipliers* of the system (1). The numbers $r_1, \ldots, r_m$ defined by the relations $\lambda_i = e^{r_i T}$, $i = 1, \ldots, m$, are called the *characteristic exponents* of the system.

We then have the following corollary to the previous theorem.

COROLLARY. *The system* (1) *has a periodic solution of period T if and only if there is at least one characteristic factor equal to unity.*

From the definition it follows that the characteristic exponents are determined up to multiples of $2\pi i/T$, where $i = \sqrt{-1}$. If λ_i is a characteristic factor of (1) and $x(t)$ is a corresponding solution satisfying $x(t + T) = \lambda_i x(t)$, then write $x(t)$ in the form

$$x(t) = p(t)e^{r_i t}.$$

Then we have

$$p(t + T)e^{r_i(t+T)} = \lambda_i p(t)e^{r_i t},$$

and by the definition of r_i, it follows that $p(t + T) = p(t)$. Thus we have the following result.

THEOREM B.2. *If $r_1, \ldots, r_m$ are the characteristic exponents of the system* (1), *then there are at least m solutions of the form*

$$x_i(t) = p_i(t)e^{r_i t}, \qquad i = 1, \ldots, m,$$

where the functions $p_i(t)$ are periodic with period T.

Finally, suppose we take multiplicities into account and denote by $\lambda_1, \ldots, \lambda_n$ the characteristic roots of the matrix C, and by $r_1, \ldots, r_n$ the corresponding exponents. If we choose the fundamental matrix $\Phi(t)$ so that $\Phi(0) = I$, then $C = \Phi(T)$. From linear algebra and the previous discussion follows the relation

$$\begin{aligned}\exp(r_1 + \cdots + r_n)T &= \lambda_1 \cdots \lambda_n = \det \Phi(T)\\ &= \det C = \exp\left[\int_0^T \operatorname{tr} A(s)\, ds\right],\end{aligned}$$

and therefore

$$r_1 + \cdots + r_n \equiv T^{-1} \int_0^T \operatorname{tr} A(s)\, ds \qquad (\text{mod } 2\pi i/T).$$

Example: Consider the second-order linear equation (Hill's equation)

$$\ddot{x} + p(t)x = 0,$$

where $p(t)$ is real-valued, continuous, and periodic with period T, say $T = \pi$. We choose a fundamental system of solutions $x_1(t)$, $x_2(t)$ satisfying

$$x_1(0) = 1, \qquad \dot{x}_1(0) = 0, \qquad x_2(0) = 0, \qquad \dot{x}_2(0) = 1,$$

and hence their Wronskian equals 1. The preceding results imply that

$$C = \begin{pmatrix} x_1(\pi) & x_2(\pi) \\ \dot{x}_1(\pi) & \dot{x}_2(\pi) \end{pmatrix},$$

since

$$A(t) = \begin{pmatrix} 0 & 1 \\ -p(t) & 0 \end{pmatrix} \qquad \text{and} \qquad \Phi(t) = \begin{pmatrix} x_1(t) & x_2(t) \\ \dot{x}_1(t) & \dot{x}_2(t) \end{pmatrix}.$$

The characteristic roots λ_1, λ_2 are therefore roots of the equation

$$\det(\lambda I - C) = \lambda^2 - [x_1(\pi) + \dot{x}_2(\pi)]\lambda + 1 = 0,$$

which implies that $\lambda_1 \lambda_2 = 1$. The characteristic exponents r_1, r_2 satisfy the relation

$$r_1 + r_2 \equiv 0 \qquad (\text{mod } 2i),$$

so we may define (up to an integer multiple of 2) the number r satisfying

$$e^{ir\pi} = \lambda_1, \qquad e^{-ir\pi} = \lambda_2 .$$

From Floquet's theorem and its corollaries it follows that

(*i*) if $\lambda_1 \neq \lambda_2$, then Hill's equation has two linearly independent solutions,

$$\varphi_1(t) = e^{irt} p_1(t), \qquad \varphi_2(t) = e^{-irt} p_2(t),$$

where $p_1(t)$ and $p_2(t)$ are periodic with period π;

(*ii*) if $\lambda_1 = \lambda_2$, then Hill's equation has a nontrivial solution, which is periodic with period π (when $\lambda_1 = \lambda_2 = 1$) or period 2π (when $\lambda_1 = \lambda_2 = -1$).

In the last case the solution is of the form $e^{i2t}p(t)$ or $e^{it}p(t)$, where $p(t)$ is periodic with period π.

The difficulty in applying the above results is that we must have information about the matrix C, or equivalently, about a fundamental system of solutions of (1), to be able to determine the characteristic exponents. But for certain systems and in particular for Hill's equation, conditions on $A(t)$ have been given that guarantee that there exist stable, unstable, or oscillatory solutions.

We conclude by giving a result for the nonhomogeneous system

$$\dot{x} = A(t)x + B(t), \tag{3}$$

where $A(t)$ and $B(t)$ are continuous and periodic with period T.

THEOREM B.3: *The system* (3) *has a periodic solution of period T for every* $B(t)$ *if and only if the corresponding homogeneous system has no nontrivial solution of period T.*

Proof. A periodic solution $x(t)$ with period T satisfies the relation $x(0) = x(T)$; conversely, if there exists a solution satisfying this relation, then it must be periodic. For if $x(t)$ is a solution, then $y(t) = x(t + T)$ is also a solution since $A(t)$ and $B(t)$ are periodic. The relation $x(0) = x(T)$ implies $x(0) = y(0)$, and by uniqueness this implies $x(t) = y(t)$ for all t; hence $x(t)$ is periodic.

By the results of Chapter 2, the solution $x(t)$ of (3) satisfying $x(0) = x_0$ is given by

$$x(t) = \Phi(t)x_0 + \int_0^t \Phi(t)\Phi^{-1}(s)B(s)\, ds, \qquad -\infty < t < \infty,$$

where $\Phi(t)$ is a fundamental matrix of the corresponding homogeneous system and $\Phi(0) = I$. From the previous remarks, the existence of a periodic solution is therefore equivalent to the relation

$$x(0) = x_0 = x(T) = \Phi(T)x_0 + \int_0^T \Phi(T)\Phi^{-1}(s)B(s)\, ds,$$

or equivalently, for every $B(t)$ we can solve

$$[I - \Phi(T)]x_0 = \int_0^T \Phi(T)\Phi^{-1}(s)B(s)\, ds.$$

But the last equation will have a solution x_0 if and only if $\det[I - \Phi(T)] \neq 0$, which is equivalent to the assertion that the equation $\Phi(T)x_0 = x_0$ has only the trivial solution $x_0 = 0$.

However, the solutions of the corresponding homogeneous system $\dot{x} = A(t)x$ are given by $\varphi(t) = \Phi(t)x_0$, and therefore the relation

$$\varphi(T) = \Phi(T)x_0 = \varphi(0) = x_0$$

can only be satisfied by the trivial solution $\varphi(t) \equiv 0$. Again by the remarks above, this is equivalent to the assertion that the homogeneous system has only the trivial solution as a periodic solution of period T, and this proves the theorem.

References

General References and Introductory Works

1. G. Birkhoff and G. Rota, *Ordinary Differential Equations*, Ginn, Boston, 1962.
2. F. Brauer and J. A. Nohel, *Ordinary Differential Equations*, Benjamin, New York, 1967.
3. J. L. Brenner, *Problems in Differential Equations*, 2nd ed., W. H. Freeman and Company, San Francisco and London, 1966.
4. E. Coddington, *An Introduction to Ordinary Differential Equations*, Prentice-Hall, Englewood Cliffs, N.J., 1961.
5. E. Coddington and N. Levinson, *Theory of Ordinary Differential Equations*, McGraw-Hill, New York, 1955.
6. W. Kaplan, *Ordinary Differential Equations*, Addison-Wesley, Reading, Mass., 1958.
7. W. Leighton, *Ordinary Differential Equations*, 2nd ed., Wadsworth, Belmont, Calif., 1966.
8. K. Yosida, *Lectures on Differential and Integral Equations*, Interscience, New York, 1960.

References Emphasizing Stability Theory

1. R. Bellman, *Stability Theory of Ordinary Differential Equations*, McGraw-Hill, New York, 1953.

2. L. Cesari, *Asymptotic Behavior and Stability Problems in Ordinary Differential Equations*, 2nd ed., Academic Press, New York, 1963.
3. W. Coppel, *Stability and Asymptotic Behavior of Differential Equations*, Heath, Boston, 1965.
4. W. Hahn, *Theory and Application of Liapunov's Direct Method* (translated by H. Hosenthien and S. Lehnigh), Prentice-Hall, Englewood Cliffs, N.J., 1963.
5. A. Halanay, *Differential Equations*, Academic Press, New York, 1966.
6. W. Hurewicz, *Lectures on Ordinary Differential Equations*, M.I.T. Press, Cambridge, Mass., 1958.
7. J. LaSalle and S. Lefschetz, *Stability by Liapunov's Direct Method with Applications*, Academic Press, New York, 1961.
8. S. Lefschetz, *Differential Equations: Geometric Theory*, 2nd ed., Interscience, New York, 1963.
9. L. Pontryagin, *Ordinary Differential Equations* (translated by L. Kacinskas and W. Counts), Addison-Wesley, Reading, Mass., 1962.

Index

A CATALOG OF SELECTED

DOVER BOOKS

IN SCIENCE AND MATHEMATICS

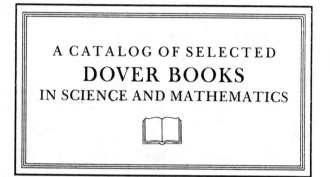

NUMERICAL METHODS FOR SCIENTISTS AND ENGINEERS, Richard Hamming. Classic text stresses frequency approach in coverage of algorithms, polynomial approximation, Fourier approximation, exponential approximation, other topics. Revised and enlarged 2nd edition. 721pp. 5⅜ × 8½.
65241-6 Pa. $14.95

THEORETICAL SOLID STATE PHYSICS, Vol. I: Perfect Lattices in Equilibrium; Vol. II: Non-Equilibrium and Disorder, William Jones and Norman H. March. Monumental reference work covers fundamental theory of equilibrium properties of perfect crystalline solids, non-equilibrium properties, defects and disordered systems. Appendices. Problems. Preface. Diagrams. Index. Bibliography. Total of 1,301pp. 5⅜ × 8½. Two volumes. Vol. I 65015-4 Pa. $14.95
Vol. II 65016-2 Pa. $14.95

OPTIMIZATION THEORY WITH APPLICATIONS, Donald A. Pierre. Broad-spectrum approach to important topic. Classical theory of minima and maxima, calculus of variations, simplex technique and linear programming, more. Many problems, examples. 640pp. 5⅜ × 8½. 65205-X Pa. $14.95

THE CONTINUUM: A Critical Examination of the Foundation of Analysis, Hermann Weyl. Classic of 20th-century foundational research deals with the conceptual problem posed by the continuum. 156pp. 5⅜ × 8½. 67982-9 Pa. $5.95

ESSAYS ON THE THEORY OF NUMBERS, Richard Dedekind. Two classic essays by great German mathematician: on the theory of irrational numbers; and on transfinite numbers and properties of natural numbers. 115pp. 5⅜ × 8½.
21010-3 Pa. $4.95

THE FUNCTIONS OF MATHEMATICAL PHYSICS, Harry Hochstadt. Comprehensive treatment of orthogonal polynomials, hypergeometric functions, Hill's equation, much more. Bibliography. Index. 322pp. 5⅜ × 8½. 65214-9 Pa. $9.95

NUMBER THEORY AND ITS HISTORY, Oystein Ore. Unusually clear, accessible introduction covers counting, properties of numbers, prime numbers, much more. Bibliography. 380pp. 5⅜ × 8½. 65620-9 Pa. $9.95

THE VARIATIONAL PRINCIPLES OF MECHANICS, Cornelius Lanczos. Graduate level coverage of calculus of variations, equations of motion, relativistic mechanics, more. First inexpensive paperbound edition of classic treatise. Index. Bibliography. 418pp. 5⅜ × 8½. 65067-7 Pa. $11.95

MATHEMATICAL TABLES AND FORMULAS, Robert D. Carmichael and Edwin R. Smith. Logarithms, sines, tangents, trig functions, powers, roots, reciprocals, exponential and hyperbolic functions, formulas and theorems. 269pp. 5⅜ × 8½. 60111-0 Pa. $6.95

THEORETICAL PHYSICS, Georg Joos, with Ira M. Freeman. Classic overview covers essential math, mechanics, electromagnetic theory, thermodynamics, quantum mechanics, nuclear physics, other topics. First paperback edition. xxiii + 885pp. 5⅜ × 8½. 65227-0 Pa. $19.95

SPECIAL FUNCTIONS, N.N. Lebedev. Translated by Richard Silverman. Famous Russian work treating more important special functions, with applications to specific problems of physics and engineering. 38 figures. 308pp. 5⅜ × 8½. 60624-4 Pa. $8.95

OBSERVATIONAL ASTRONOMY FOR AMATEURS, J.B. Sidgwick. Mine of useful data for observation of sun, moon, planets, asteroids, aurorae, meteors, comets, variables, binaries, etc. 39 illustrations. 384pp. 5⅜ × 8¼. (Available in U.S. only) 24033-9 Pa. $8.95

INTEGRAL EQUATIONS, F.G. Tricomi. Authoritative, well-written treatment of extremely useful mathematical tool with wide applications. Volterra Equations, Fredholm Equations, much more. Advanced undergraduate to graduate level. Exercises. Bibliography. 238pp. 5⅜ × 8½. 64828-1 Pa. $7.95

POPULAR LECTURES ON MATHEMATICAL LOGIC, Hao Wang. Noted logician's lucid treatment of historical developments, set theory, model theory, recursion theory and constructivism, proof theory, more. 3 appendixes. Bibliography. 1981 edition. ix + 283pp. 5⅜ × 8½. 67632-3 Pa. $8.95

MODERN NONLINEAR EQUATIONS, Thomas L. Saaty. Emphasizes practical solution of problems; covers seven types of equations. ". . . a welcome contribution to the existing literature. . . ."—*Math Reviews*. 490pp. 5⅜ × 8½. 64232-1 Pa. $11.95

FUNDAMENTALS OF ASTRODYNAMICS, Roger Bate et al. Modern approach developed by U.S. Air Force Academy. Designed as a first course. Problems, exercises. Numerous illustrations. 455pp. 5⅜ × 8½. 60061-0 Pa. $9.95

INTRODUCTION TO LINEAR ALGEBRA AND DIFFERENTIAL EQUATIONS, John W. Dettman. Excellent text covers complex numbers, determinants, orthonormal bases, Laplace transforms, much more. Exercises with solutions. Undergraduate level. 416pp. 5⅜ × 8½. 65191-6 Pa. $10.95

INCOMPRESSIBLE AERODYNAMICS, edited by Bryan Thwaites. Covers theoretical and experimental treatment of the uniform flow of air and viscous fluids past two-dimensional aerofoils and three-dimensional wings; many other topics. 654pp. 5⅜ × 8½. 65465-6 Pa. $16.95

INTRODUCTION TO DIFFERENCE EQUATIONS, Samuel Goldberg. Exceptionally clear exposition of important discipline with applications to sociology, psychology, economics. Many illustrative examples; over 250 problems. 260pp. 5⅜ × 8½. 65084-7 Pa. $7.95

LAMINAR BOUNDARY LAYERS, edited by L. Rosenhead. Engineering classic covers steady boundary layers in two- and three-dimensional flow, unsteady boundary layers, stability, observational techniques, much more. 708pp. 5⅜ × 8½. 65646-2 Pa. $18.95

LECTURES ON CLASSICAL DIFFERENTIAL GEOMETRY, Second Edition, Dirk J. Struik. Excellent brief introduction covers curves, theory of surfaces, fundamental equations, geometry on a surface, conformal mapping, other topics. Problems. 240pp. 5⅜ × 8½. 65609-8 Pa. $8.95

CATALOG OF DOVER BOOKS

CHALLENGING MATHEMATICAL PROBLEMS WITH ELEMENTARY SOLUTIONS, A.M. Yaglom and I.M. Yaglom. Over 170 challenging problems on probability theory, combinatorial analysis, points and lines, topology, convex polygons, many other topics. Solutions. Total of 445pp. 5⅜ × 8½. Two-vol. set.
Vol. I 65536-9 Pa. $7.95
Vol. II 65537-7 Pa. $6.95

FIFTY CHALLENGING PROBLEMS IN PROBABILITY WITH SOLUTIONS, Frederick Mosteller. Remarkable puzzlers, graded in difficulty, illustrate elementary and advanced aspects of probability. Detailed solutions. 88pp. 5⅜ × 8½.
65355-2 Pa. $4.95

EXPERIMENTS IN TOPOLOGY, Stephen Barr. Classic, lively explanation of one of the byways of mathematics. Klein bottles, Moebius strips, projective planes, map coloring, problem of the Koenigsberg bridges, much more, described with clarity and wit. 43 figures. 210pp. 5⅜ × 8½. 25933-1 Pa. $5.95

RELATIVITY IN ILLUSTRATIONS, Jacob T. Schwartz. Clear nontechnical treatment makes relativity more accessible than ever before. Over 60 drawings illustrate concepts more clearly than text alone. Only high school geometry needed. Bibliography. 128pp. 6⅛ × 9¼. 25965-X Pa. $6.95

AN INTRODUCTION TO ORDINARY DIFFERENTIAL EQUATIONS, Earl A. Coddington. A thorough and systematic first course in elementary differential equations for undergraduates in mathematics and science, with many exercises and problems (with answers). Index. 304pp. 5⅜ × 8½. 65942-9 Pa. $8.95

FOURIER SERIES AND ORTHOGONAL FUNCTIONS, Harry F. Davis. An incisive text combining theory and practical example to introduce Fourier series, orthogonal functions and applications of the Fourier method to boundary-value problems. 570 exercises. Answers and notes. 416pp. 5⅜ × 8½. 65973-9 Pa. $9.95

THE THEORY OF BRANCHING PROCESSES, Theodore E. Harris. First systematic, comprehensive treatment of branching (i.e. multiplicative) processes and their applications. Galton-Watson model, Markov branching processes, electron-photon cascade, many other topics. Rigorous proofs. Bibliography. 240pp. 5⅜ × 8½. 65952-6 Pa. $6.95

AN INTRODUCTION TO ALGEBRAIC STRUCTURES, Joseph Landin. Superb self-contained text covers "abstract algebra": sets and numbers, theory of groups, theory of rings, much more. Numerous well-chosen examples, exercises. 247pp. 5⅜ × 8½. 65940-2 Pa. $7.95

Prices subject to change without notice.

Available at your book dealer or write for free Mathematics and Science Catalog to Dept. GI, Dover Publications, Inc., 31 East 2nd St., Mineola, N.Y. 11501. Dover publishes more than 175 books each year on science, elementary and advanced mathematics, biology, music, art, literature, history, social sciences and other areas.